DO YOU KNOW . . .

- What creature can broad-jump 300 times its own length?
- What is the only domestic animal not mentioned in the King James Bible?
- What sea animal turns red, white, blue, and green according to its mood?
- What is the only mammal that can truly fly?
- What animal has the greatest range of vision?

The answers are all here—and much, much more.

Joan Embery invites you to discover a treasure of little-known lore about mammals, fish, birds, reptiles, and insects. She has put all that she knows and loves about animals into this book for animal lovers, list fanatics, trivia freaks, and her very own fans. This marvelous collection is filled with over 1,500 incredible, amazing, fun, interesting, and important facts to expand your knowledge and to enrich your love and understanding of the astonishing animal kingdom.

Joan Embery's Collection of Amazing Animal Facts

4.bat 5.giraffe

D0424567

Books by Joan Embery

MY WILD WORLD (with Denise Demong)

Joan Embery's Collection of Amazing Animal Facts

JOAN EMBERY

with Ed Lucaire

with illustrations by Karel Havlicek

A DELL BOOK

Steve Joines helped greatly with the research.
My thanks to him for his knowledge and enthusiasm.

Published by
Dell Publishing Co., Inc.
1 Dag Hammarskjold Plaza
New York, New York 10017

First Dell printing—February 1984

Introduction

After writing my first book, *My Wild World*, in which I described the marvelous experiences I've had as goodwill ambassador for the San Diego Zoo, I knew I wanted to write another book about animals. I thought of several book ideas, but I finally settled on doing this encyclopedia because animals are fascinating even when you don't know too much about them and, naturally, the more you learn, the more interesting animals become and the more pleasure you can get from watching and reading about them.

I wanted this book to be more than just a compilation of facts that would expand your knowledge. I wanted it to be fun too. So I went through my files looking for facts that weren't too widely known and were amusing, odd, bizarre, cute, and even puzzling. I selected hundreds of facts and then went through them several times to determine which were the most entertaining. This collection is the result.

I enjoyed doing the research. For me, it wasn't really work, but then I've always been that way: if whatever I was doing had something to do with animals, I loved it. I can't remember a time when I wasn't crazy about animals. When I was a kid, I kept any pet I could talk my parents into letting me have. They were understanding, though they must have felt overrun sometimes by my ducks, rabbits, dogs, guinea pigs, and chickens. I even tamed squirrels, possums, and snakes. My favorite animal, though, wasn't a household pet. I was absolutely mad about horses. Still am.

I was the most hypnotized by animals, but my family enjoyed them too. I was fortunate in having two family members who worked with animals professionally. My mother was a vet's assistant for many years, and my uncle was and still is a veterinarian. I suppose it was almost inevitable that when I was very young I wanted to be a vet. I was lucky enough to spend every summer with my uncle when I was growing up. On those visits my twin sister, Linda, and I would always haunt his veterinary hospital.

Through high school my ambition to be a vet didn't change, and I'd probably be one now if a happy opportunity hadn't occurred. I'd enrolled at San Diego State University as a pre-vet major. I was already looking ahead to veterinary school and, because I'd been told that practical experience was an important factor in gaining admission, I'd decided to get a part-time job working with animals. At the end of my freshman year I found what I'd been looking for: attendant at the Children's Zoo of the San Diego Zoo, one of the best zoos in the world.

My job was such a joy to me that after a while my priorities changed and when, after less than two years, I was selected as the San Diego Zoo's goodwill ambassador, I became a part-time student. Being the Zoo's ambassador was a tremendous opportunity for me. It meant giving lectures and making television appearances with special zoo animals to promote the Zoo. I wanted to be as thoroughly prepared as possible for my lectures and TV appearances, and it was then that I began doing such extensive research and building such voluminous files that I was able to compile this book from my files alone.

In the course of my lectures and appearances I have tried to make my animals and myself entertaining, but I have also tried to inform and alert people about a very serious matter: the threat of extinction that too many animal species face. There are now approximately 1,000 species on the endangered list. Animals enrich our lives and they should be saved. Please do what you can for the world's animals.

Joan Embery
San Diego Zoo
August 1982

Aardvark

In order to escape predators, aardvarks are able to burrow into the ground at incredibly high rates of speed.

The aardvark's sticky tongue can be extended as far as 12 inches, an adaptation for reaching and capturing the ants and termites on which it feeds.

The aardvark's relationship to other mammals has always been a source of debate among zoologists. Aardvarks have no immediate ancestors but probably arose from very early hoofed mammals.

The word *aardvark* is Afrikaans meaning "earth pig," which is what the Dutch colonists of South Africa thought the animal looked like.

Aardvarks are the only members of the zoological order Tubulidentata, which literally translates as "tube tooth" because the teeth are simple cylinders of dentine traversed from the base to crown with hundreds of minute passages.

Aardvarks lack front teeth, and their cheek teeth have no enamel and grow continuously.

An aardvark can close its nostrils in order to keep dirt and termites out of its nose. It can also fold its ears back to keep dirt out of them while burrowing.

Not a small animal, the average aardvark is about 6 feet in length and weighs 180 pounds. Aardvarks are hunted as food by South African natives.

In zoos aardvarks (and anteaters) are fed milk, ground meat, and vitamins, because adequate supplies of ants and termites would be difficult to provide.

Aardwolf

Aardwolves are small animals, related to the hyena, which inhabit eastern and southern Africa. The name *aardwolf* is Afrikaans, meaning "earth wolf." The animals are so named because they live in burrows and feed mostly on termite nests.

Aardwolves may be a case of true mammalian mimicry. Small and nonaggressive animals, an aardwolf's coat is patterned similarly to that of the large, fierce striped hyena.

Abalone

Abalone, a popular delicacy in California and Australia, is a type of sea snail. The edible part of the snail is the large foot.

Abalone occasionally produce beautiful iridescent pearls.

Agouti

The agouti, a South American rodent measuring about 20 inches and weighing about 6 pounds, is capable of jumping 6½ feet into the air from a standing start.

Albatross

The thirteen species of albatross have some of the longest incubation periods of any egg-laying animals. The single egg is incubated for up to 70 days with both parents alternately sitting on the nest. After hatching, chicks remain in the nest for almost 11 months.

Studies have shown that parent birds who are feeding a chick may stray as far as 2,500 miles from the nest site in search of food.

Like storks, albatross click their beaks rapidly during their elaborate premating ritual.

After a young albatross is fledged and leaves the nest, it may go to sea for an extended period of time, never touching land.

The wingspan of the wandering albatross (Diomedea exulans) is the largest of any living bird, averaging just over 10 feet, although the largest ever recorded measured 11 feet 10 inches. The long wings of the albatross are superbly adapted for gliding on the air currents of the southern oceans.

Albatross are able to drink seawater without harm because a specialized nasal gland quickly discharges excess salt from their systems.

The word albatross is an English corruption of "alcatraz," the name that Portuguese seamen applied to pelicans and all large seabirds.

Because of their weight and long wings, albatross require a wind in order to become airborne; they cannot take flight in calm weather.

Albatross have virtually no enemies when they are at sea and usually live long lives. They do not attain sexual maturity until they are seven to ten years of age, longer than many birds live.

Alligator

The muscles that are used to close an alligator's jaw are extremely powerful, an adaptation that aids these reptiles in subduing their prey. The muscles that are used for opening the jaw, however, are weak, which makes it easy to hold an alligator's jaw shut.

After depositing her eggs in a nest that she has constructed, a female alligator will keep guard over them during incubation.

Outside the United States, alligators are found only in one area of the world, in the Yangtze River of China and its tributaries. The only location where alligators and crocodiles coexist is in southern Florida.

The word *alligator* is derived from the Spanish noun *el lagarto,* meaning "lizard."

Alligators can be distinguished from crocodiles by their more rounded snout and by the fact that their teeth are not visible when the mouth is closed.

Alpaca

The alpaca (and the llama) is a domestic form of the guanaco, a humpless South American relative of the camels.

Camels evolved in the New World. The camels of the Old World migrated to their present ranges via the Bering land bridge.

The wool of the alpaca is extremely fine, being one half the diameter of cashmere goat hair and one quarter the diameter of a human hair.

Amphibian

The word *amphibia* means "double life" and refers to the ability of these animals to live in the water or on the land.

Amphibians are adapted for life in fresh water only, since even a small amount of salt is fatal to tadpoles.

An exception to this rule is the Argentinian toad *(Bufo arenorum),* which breeds and deposits its eggs near the coast in pools of brackish water.

Frogs are the widest-ranging amphibians, being found on every continent except Antarctica.

Amphibians are distinguished from reptiles by their naked, moist skin, which is used in respiration. The skin has a large number of capillaries, supplying blood that picks oxygen up directly from the air. Some species gain as much as 70% of their oxygen through skin respiration.

The earliest known amphibian and the earth's earliest known quadruped was *Ishthyostega*. This animal inhabited Greenland around 350 million years ago.

The world's largest amphibian is the Chinese giant salamander, which averages 40 inches in length and weighs about 28 pounds. The largest giant salamander ever recorded measured 5 feet and weighed 100 pounds.

There are some 2,000 living species of amphibians divided into 3 orders. The largest order is Salientia, the frogs and toads, containing about 1,800 species.

Anchovy

Anchovies are indirectly responsible for some of the world's best fertilizer. On islands off the coast of Peru millions of boobies and cormorants feed on anchovies and ultimately produce the fertilizer known as guano.

Angler Fish

The first ray of the angler fish's dorsal fin acts as a fishing pole and lure, the end of which looks like a small fish's tail. The lure attracts prey close enough for the angler fish to open its mouth and swallow the fish.

Angler fish move along the ocean floor on paired fins that function like legs.

Angler fish can attain a length of five feet.

Ant

Scientists estimate that there are one quadrillion (1,000,000,000,000,000) ants living on the earth at any given time.

An ant can lift and carry objects that are more than 50 times its own weight.

For hundreds of years native peoples in South America and Africa have used the pincers of soldier and driver ants for suturing wounds because they are working ants with exceptionally large heads and jaws. The ants are placed on the wound and allowed to bite, forming the suture; then their bodies are cut off, leaving the mandibles locked in position, holding the skin.

Parasol ants have the curious habit of cutting up small pieces of leaves and carrying them on their heads to their colony. There they deposit the cuttings in underground chambers to let them spawn fungus on which the parasol ants feed.

Most driver and soldier ants are blind and seek out with their sense of smell the animals on which the colony feeds.

Weaver ants of Africa and Australia make their nests in trees by selecting large leaves and collectively folding them over and sealing the edges with an adhesive produced by their larvae.

Honeypot ants, which store a honeylike substance in their abdomens, are eaten by natives of Australia and Central America.

Anteater

There are several different species of animal that are popularly referred to as anteaters. Among these are the strange, primitive spiny anteater of Australia and New Guinea, which is a monotreme (egg-laying mammal) like the duck-billed platypus; the giant anteater of South America; and the largely arboreal tamandua.

With the exception of the spiny anteaters, anteaters are placed in the order Edentata (meaning without teeth) along with the sloths and armadillos; only the anteaters, however, truly lack teeth.

The giant anteater ranges in length from 5 feet to almost 7 feet and weighs an average of 50 pounds. It possesses 3 enormous (up to 6 inches) claws on its front feet, which it uses to rip open the nests of ants and termites, eating the eggs, cocoons and adults that it catches with its sticky saliva-coated tongue.

The salivary glands of the giant anteater appear to function only when the animal is feeding. Its tongue can be extended as far as 2 feet but is only 10–15 mm (⅝″) in diameter.

Giant anteaters are normally solitary except for females with young. A young anteater is carried around on its mother's back. Anteaters are good swimmers and are known to cross wide rivers. The word *tamandua* is Brazilian, meaning "ant trap."

Antelope

The word *antelope* was first used to describe the Indian black buck. The word is derived from the Greek *antholps,* meaning "brightness of eye."

There are some 85 species of antelope. Antelope are members of the family Bovidae, even-toed, hollow-horned ruminants (cud-chewers), also containing buffaloes, bison, cattle, sheep, and goats.

Antelope have no upper incisor teeth but rather a tough, horny pad that meets the lower incisors to clip vegetation. True antelope are not found in the New World. The pronghorn, often referred to as an antelope, is placed in a family by itself, the Antilocapridae. Pronghorns have branched horns whose outer leathery sheath is shed and renewed annually, giving them some of the characteristics of deer.

The difference between horns and antlers is that horns are permanent, hollow, unbranched, and are supported by bony cores projecting from the skull, while antlers are branched, grown and shed annually, and are not permanently attached to the skull.

The royal antelope of equatorial Africa, standing less than a foot high at the shoulder, is the smallest antelope and the world's smallest ruminant. Even so, the royal antelope has been recorded clearing an eight-foot barrier in a single leap.

giant eland

The cape eland, one of the largest of the antelope, has also been recorded leaping over eight-foot barriers, which is a remarkable feat for an animal weighing 1,300–1,400 pounds.

Cape eland are easily tamed and are more easily fattened than other antelope species. Their flesh and milk are excellent, and they have been kept in semicaptivity and semidomestication in South Africa, Rhodesia, and Askaniya-Nova, U.S.S.R., for more than 60 years.

The largest of all antelope is the western giant eland or Lord Derby's eland, which is now extremely rare and confined to one national park in Senegal and to a small scattered population surviving along the border between Senegal and Mali, where they are endangered due to poaching. The western giant eland stands over six feet high at the shoulder.

Ape

Apes are members of the family Pongidae, which contains five genera and about ten species. The great apes include the gorilla, orangutan, and chimpanzee; the lesser apes are the gibbons and the siamang.

It is often said that the way to tell the difference between an ape and a monkey is that monkeys have tails and apes do not. This is not true, however, since several species of monkey have very short tails or none at all.

The great apes generally support themselves on all four limbs in a semierect posture but all of them are able to walk on their legs only.

The apes are humankind's nearest living relatives.

In general form and structure the brain of the ape closely resembles the human brain.

Apes are susceptible to all the same diseases as humans and great care must be taken with captive apes to ensure their health. Apes are especially susceptible to respiratory ailments in captivity.

Of all the mammals gibbons are the most agile tree dwellers.

Archaeopteryx

Archaeopteryx is the earliest known bird, dating back to around 140 million years ago. The first fossil *Archaeopteryx* was found in a slate quarry at Langeraltheim, Bavaria, in 1861, and would have been classified as a reptile except for the clear imprint of feathers that had been preserved in the fine grains of the slate-producing sediments.

Archaeopteryx means "ancient wing."

To date, only three fossils of *Archaeopteryx* have been discovered, the original in 1861, another discovered ten miles away in 1877, and a third unearthed in 1956.

Archaeopteryx represents a true "missing link" between the reptiles and birds.

The wing bones of *Archaeopteryx* terminated in three slender clawed fingers. From an analysis of its skeleton *Archaeopteryx* probably could not fly but rather moved about on the ground on its sturdy legs, climbed into trees by using its clawed wings, and, when threatened, launched itself into the air to glide away much like modern flying squirrels.

Arctic Tern

The arctic tern migrates the greatest distance of any bird species, from the Arctic to the Antarctic and back, the equivalent of a flight around the world.

Armadillo

A female armadillo usually gives birth to four genetically identical infants of the same sex.

The armadillo is the only animal other than humans that is known to contract leprosy; for this reason they are used in laboratories to study the disease.

Although the armadillo is a member of the order Edentata (toothless mammals), it has up to 100 teeth.

Aurochs

Aurochs were a species of wild cattle that inhabited Europe and were the progenitors of many of our domestic varieties of cattle.

The last known aurochs was speared to death in 1627.

Aye-Aye

The aye-aye, one of the world's most endangered animals, is a primate inhabiting only the island of Madagascar. Aye-ayes are the size of large domestic cats and when first discovered by science were thought to be rodents rather than primates because their chisellike, constantly growing incisor teeth are very much like those of the rodents.

Aye-ayes are nocturnal and feed almost entirely on insects living in dead tree limbs and under bark. They locate insects by listening for their movements and capture them with a specialized, extremely elongated middle finger that they use to probe under the bark.

Aye-ayes use their elongated middle finger for a variety of tasks other than catching insects, including grooming and drinking. When drinking, the aye-aye dips all its fingers into water, then draws them quickly through its mouth.

The aye-aye also uses its long finger to flick other liquids like the contents of an egg into its mouth. They do this very rapidly, the finger a blur.

Malagasy natives believe that aye-ayes are the spirits of their departed ancestors.

In German the name for the aye-aye is *Fingertier* or "finger animal."

Baboon

There are eight species of baboons ranging throughout Africa and the Arabian Peninsula. Of these eight, seven are members of the genus *Papio*. The remaining species is the gelada *(Theropithicus gelada)*, which inhabits rocky areas of Ethiopia at high elevations.

The hamadryas baboon was the sacred baboon of the ancient Egyptians. They were believed to be representatives of the god Thoth, god of letters and scribe of the gods. It was believed that Thoth had learned the sacred hieroglyphics from baboons and by Egyptian tradition: baboons were said to be able to write.

Hamadryas baboons were mummified, entombed, and associated with sun worship by the ancient Egyptians.

Baboons normally live in troops of 20–50 individuals (though troops of up to 300 have been reported) of all ages, who are led by one or more adult males. The social structure of the troop is well defined and rigidly maintained by the dominant males. In part, this rigid social structure functions to protect the troop from enemies such as leopards.

Normally when a baboon troop is foraging for food, the troop moves in a defensive formation with young males foraging at the front and rear of the group while the females and infants stay toward the center near the dominant males. When danger threatens, the dominant males charge toward the source of trouble and lead the other males in action against the predator while the females and young make their escape.

Adult male baboons are large, powerful, and aggressive animals well equipped for dealing with predators, especially when they work as a unit. The canine teeth of an adult male baboon are larger than those of a leopard.

Baboons threaten other animals and one another by yawning to expose their massive canine teeth.

Gelada baboons have an even more impressive threat gesture known as the "lip-flip." They draw up the upper eyelids to expose the light-colored skin surrounding the eyes, then "flip" their lips in such a way that the upper lip covers the nose and the lower lip covers the chin, thereby exposing their canines in a grotesque, threatening face.

Bacterium

A new species of bacterium was discovered in 1973 in San Francisco's famous sourdough bread. It was given the name *Lactobacillus san franciscoi* and is responsible for giving the bread its sour taste.

Badger

The German word for "badger" is *Dachs* and the dachshund earned its name because it was bred to have legs short enough to crawl into badger burrows and chase the animals out.

Badger fur was once widely used in the manufacture of shaving brushes.

Barnacle

Barnacles are crustaceans related to lobsters, crabs, and shrimp. Despite the presence of a mollusklike shell as an adult, young barnacles are mobile and resemble other crustaceans. After finding a site on which to settle, however, they anchor themselves and secrete a calcium compound forming the familiar barnacle shell.

Basenji

The Basenji dog of Africa does not bark and is therefore a favored hunting dog.

Bass

The largemouth bass of the United States and Canada is the largest member of the sunfish family.

Bat

Among mammals, only rodents exceed bats in total

number of species. The order is divided into 18 families and 180 genera.

Despite the phrase "blind as a bat," the fruit bats of the Old World have excellent vision. Many bats do, however, have weak vision and compensate by using their highly specialized sense of echolocation. Bats emit ultrasonic clicks that bounce off solid objects and alert the bats to their presence.

Bats are the only mammals that truly fly. All other "flying" mammals are actually gliders.

Bats sleep upside down because this position makes it easy for them to take wing quickly should danger threaten. They simply let go and spread their wings to escape to safety.

In flight, bats utilize both their arms and legs, actually "swimming" through the air.

Bats require more water and moist food than other mammals because their wings provide such great evaporative surfaces in comparison to their weight.

Most species of bats have two teats. Only humans, monkeys, apes, elephants, and a few other animals exhibit this characteristic.

Most bats produce only one offspring per year.

In Germany bats are called *Fledermaus,* or "flying mouse."

Bats are omens of good fortune in China. The symbol for *fu* (bat) is the same as the symbol for happiness. The Chinese wear pendants called *wu fu,* which show five bats, wing to wing, in a circle.

For small mammals, bats are relatively long-lived. Some bats are known to have lived as long as 24 years.

The Brazilian free-tailed bat *(Tadarida brasiliensis)* is one of the most common mammals in the United States, with an estimated population of over 100 million.

Bats are extremely beneficial predators. The average bat can consume as many as 650 mosquitoes each night in addition to other insects.

A young bat clings to its mother's fur during flight.

Discovered in 1973, Kitti's hog-nosed bat *(Craseonycteris thonglongyni)* of southern Thailand is the world's smallest known mammal, with a wingspan of 160 mm (6.29 in.), body length of 29–33 mm (1.14–1.29 in.), and a weight of 1.75–2.0 gr (.062–.071 oz.). This bat, known also as the "bumblebee bat," is found in only two caves in Thailand and is the basis for a new zoological family and genus.

The flying fox of Southeast Asia is the largest bat, with a body length of 15¾ inches, a weight of up to 32 ounces, and a wingspan of up to 5 feet 7 inches.

Unlike most bats, which tend to be dark brown or black, the painted bat is brightly colored with red, black, and silver markings.

Most bats eat insects, small animals, and fruit, but the bulldog or fishing bat of South America catches small fish with its claws as it skims the surfaces of lakes and rivers.

In certain bat caves a beetle lives off droppings and produces ammonia gas as a waste product. The ammonia level in bat caves can be as high as three times that which humans can safely breathe.

Bears

In general, bears are solitary animals except during the brief mating season and during the time that a mother is rearing her cubs.

The Kodiak bear (a type of brown bear) is the largest living terrestrial carnivore. Males average 8 feet in length, stand 52 inches high at the shoulder, and weigh over 1,000 pounds.

Bears do not truly hibernate; rather, in northern climates bears den up and enter a deep sleep during the winter. Their body temperature, heart rate, and breathing do not drop drastically as in true hibernation, however. Tropical bears and polar bears do not exhibit this behavior, remaining active year round.

The Arctic was named after the Greek word for bear, *arktos,* not because the polar bear is found there, but in reference to the northern constellation of the bear.

The liver of the polar bear contains so much vitamin A that it is poisonous and not eaten by Eskimos.

Polar bears are a host for the parasite *Trichinella.*

The guard hairs of the polar bear's coat are hollow, providing efficient insulation. In captivity polar bears may have a greenish tinge during warm months because common algae invade and grow inside these hollow hairs. In addition, a polar bear's hairs are actually clear, causing the animals to appear white.

Polar bears have been known to push a small block of ice in front of them while stalking seals in order to hide their black noses.

At birth a normal bear cub weighs only 8–12 ounces.

The forefeet of all bears are turned slightly inward, resulting in their characteristic shuffling gait.

Polar bears are found only in the northern Arctic. They do not inhabit the Antarctic.

The state animal of California is the extinct California grizzly.

Polar bears spend the majority of their lives on ice floes, often hundreds of miles from land. Their scientific name, *Thalarctos maritimus,* is from the Greek, meaning "sea bear."

Besides humans, the only animals that polar bears fear are killer whales and walrus.

Circus trainers consider bears to be more dangerous than large cats, because they are powerful and unpredictable, something to keep in mind in case you're ever tempted to feed a "tame" bear in a national park.

The sloth bears of India and Sri Lanka have an unusual adaptation for feeding. Their lips and snout are extremely mobile; they can close their noses at will; they lack the inner pair of upper incisor teeth, which forms a gap, and the palate is hollowed. These features enable sloth bears to feed on termites by digging up the nest, blowing dirt away, and sucking up termites like a vacuum.

Among American Indians killing a grizzly bear was a sign of great hunting prowess and bear-claw necklaces were worn as symbols of strength.

Beaver

Beavers are aquatic rodents. Their diet consists primarily of bark and young plant shoots.

Beavers of both sexes produce a musky secretion called castoreum. This secretion is utilized in the manufacture of perfumes.

Beavers have been known to fell trees as large as five feet in diameter. Beavers build dams in order to create pools of water that are sufficiently deep that they will not freeze to the bottom during winter. In these pools they also construct lodges with openings underwater that protect them from predators.

The rear feet of beavers have a double claw on the second toe that is used to groom their thick fur.

Beavers are monogamous and mate for life.

Beavers are occasionally killed by falling trees that they have felled.

The incisor teeth of beavers, which are orange-colored in front, continue to grow in a curve and must be used continuously to keep them at manageable length.

Bee

More people die annually from bee stings than from snake bites.

The smallest species of bee is only one third of an inch in length.

A genus of European bee is noteworthy because it makes its nest in the abandoned shells of snails.

Worker bees are all females with nonfunctioning reproductive organs; they are the product of fertilized eggs. Drones are male bees whose only function is to mate with the queen bee. Drones are the product of unfertilized eggs (parthenogenesis).

According to the principles of aerodynamics, bees should be incapable of flight because of their high bodyweight to wing area ratio.

A bee can lift or pull 300 times its own weight.

The honeycombs of giant bees in India can be as high as 7 feet and weigh as much as 400 pounds.

Bees have five eyes, three small ones on the top of the head and two large eyes on the side. Bees are able to see X rays and ultraviolet color.

Honeybees beat their wings an average of 250 times per second, almost twice as fast as the bumblebee's 130 times per second.

A queen bee can lay 1,500–3,000 eggs per day or her own weight in eggs every 24 hours.

A colony of honeybees is composed of as many as 80,000–90,000 individuals.

It is well known that the cells of a beehive are hexagonal. The hexagonal shape is extremely efficient, providing the maximum amount of enclosed space while utilizing the minimum amount of beeswax. The cells used for raising workers number five to the inch, while those used for raising drones number four to the inch.

A worker bee can collect and return to the hive with half its own weight in nectar every day.

Because honey is already digested by a bee in a special stomach used solely for that purpose, it does not have to be digested when eaten by other animals.

Australian aborigines have been known to locate beehives by capturing a bee, attaching a feather to it, and following the feather as the bee returned to the hive.

Bees and wasps are deaf.

During a good season a single hive of bees can produce up to two pounds of honey a day, requiring approximately 5 million individual bee trips to and from flowers, about 80,000 loads of pollen, and some 200,000 miles of flight.

Beetle

With perhaps as many as one million species of beetles, it is the largest single order in the animal kingdom.

The larvae of beetles are highly nutritious and are commonly eaten in parts of South America, Africa, and Australia. Beetle larvae were also considered delicacies by the ancient Greeks and Romans.

Bird

Birds outnumber all other vertebrate animals except fish.

Birds are found in virtually every habitat on earth. The only area uninhabited by birdlife is the center of the Antarctic continent.

All birds have feathers, a characteristic shared by no other type of animal.

Bird feathers have evolved from the scales of reptiles. Other reptilian features shared by the birds are the "egg teeth" possessed by the chicks of many species and the scales on birds' feet.

Birds began to evolve independently from reptilian stock around 150 million years ago, shortly after the appearance of the first mammals. Over a century ago the great naturalist Thomas Henry Huxley referred to birds as "glorified reptiles."

Ornithologists recognize 27 orders of birds, including 155 separate families and approximately 8,580 living species.

Bird species that migrate long distances may actually double their body weight in an effort to store energy up for their long journey.

Birds have the ability to sleep while perched on branches because they have special locking mechanisms in their feet.

The bones of birds are laced with air pockets to make them lighter, an adaptation for flight.

Although early forms of birds had teeth, existing birds do not. The mergansers, highly specialized fishing ducks, do have teethlike serrations on their bills that aid them in capturing fish.

The heaviest flying bird is the kori bustard of East and South Africa. It can weigh over 35 pounds and has a wingspan of 8 feet.

In general, birds have about twice as many red corpuscles per cc of blood as do mammals. This adaptation helps them to transfer oxygen to muscle tissue during the heavy stress of flight.

The largest bird that ever lived was *Aepyornis*, which may have survived on the island of Madagascar as late as 1649. The elephant bird is estimated to have weighed as much as one-half ton.

Shorebirds that nest on high sea cliffs lay eggs that taper more quickly than other bird eggs. This adaptation causes the eggs to roll in tight circles and lessens the chance of their rolling off the cliffs.

Bird of Paradise

The brightly colored exotic birds of paradise of New Guinea and Australia are relatives of the starling.

The birds of paradise have some of the most elaborate and complex courtship displays in the animal kingdom, wherein the male displays its bright plumage in an effort to attract and mate with as many females as possible.

In some species of birds of paradise the males perform their courtship displays on the wing in vertical shafts of sunlight. Incredibly, the males manufacture these light shafts themselves by laboriously stripping the leaves from a section of the forest canopy.

All of the highly ornamented species breed polygamously, and males play no part whatever in nesting activities.

Bison

Bison are the largest mammals inhabiting North America.

It is estimated that at the time of the arrival of the white man the bison population of North America stood in excess of 50 million individuals; "the greatest aggregation of large animals ever known to civilized man," according to *Encyclopaedia Britannica.* By 1889, the year when market hunters ceased hunting them, only 541 individuals remained alive. Today the American bison population is between 25,000 and 30,000.

The bison of the Wichita Mountains Preserve are the descendants of 15 animals shipped there from the Bronx Zoo in 1907 after President Theodore Roosevelt declared the area a game preserve.

Some tribes of American Indians were so dependent on the bison that their entire cultures were based on following and hunting the herds.

The European bison or wisent *(Bison bonasus)* now numbers only a few hundred individuals protected in herds in Poland and the U.S.S.R.

Cape buffalo

The word *buff* arose during the 1830s when volunteer firemen wore jackets made from bison hides. The people who followed the firemen became known as "buffs," a term subsequently used to mean any enthusiast.

A hybrid of an American bison and beef cattle called a beefalo is a promising food source of the future. The beefalo converts grass and roughage into meat better than current strains of cattle.

Bitterling Fish

The female bitterling deposits eggs into the gill chambers of freshwater mussels, after which the male deposits its milt nearby. The mussel, during the normal process of feeding, draws in the milt and unwittingly fertilizes the bitterling's eggs.

Bloodhound

The bloodhound is so named because it was the first breed of dog whose "blood" or lineage was kept track of. The record keepers were the monks of St. Hubert's Abbey. Record keeping began in the ninth century.

The bloodhound is the only animal whose "testimony" can be used as evidence in American courts.

Boa Constrictor

The oldest known snake was a boa constrictor that lived for over 40 years in the Philadelphia Zoological Gardens.

Sizes of boa constrictors tend to be exaggerated. The longest boa on record was 18 feet 2 inches.

Bobcat

Bobcats are distributed from southern Canada throughout the entire United States (except the midwestern corn belt) and into Mexico.

bobcat

Bobcats are distinguished from the closely related lynx in part because they lack the lynx's prominent ear tufts.

Like polar bears, bobcats have furred feet to aid them in moving over snow.

Boll Weevil

The town of Enterprise, Alabama, erected a monument in honor of the conquered oppressor of the cotton fields, the boll weevil, on December 11, 1919.

Bowerbird

Bowerbirds are the only bird species that build and decorate elaborate dance pavilions for courtship. When naturalists first encountered these pavilions, they believed that they were the work of humans rather than of birds.

The pavilions of the bowerbirds, who are closely related to the birds of paradise, appear to function during courtship in place of the bright plumage of their relatives.

Bowers are variable in architectural form, ranging from mats, avenued chambers, and planted lawns to teepee-roofed huts. All bowers are decorated with items such as flowers, stones, iridescent insect skeletons, and numerous kinds of berries and shells.

In some species of bowerbirds the males have brightly colored head plumes and occasionally bright body plumage, but as a general rule the better "architects" are the species with drabber plumage.

Some bowerbirds actually "paint" their bowers. They crush berries and other pigments, mix them, and apply the mixture to the walls of their bowers with a brushlike tool.

Male bowerbirds mate with as many females as they are able to attract and, like the birds of paradise, do not aid in nesting.

Bumblebee

Bumblebees have been observed flying at altitudes of 20,000 feet.

Unlike the hives of honeybees, bumblebee hives last for one season and before winter all members of the colony die, with the exception of the queen, who hibernates during the winter.

Butterfly

The African butterfly, *Druryeia antimachus,* with a large ten-inch wingspan, is so poisonous that if eaten it could be fatal to a large animal.

Every year millions of monarch butterflies migrate from Alaska and Canada to the town of Pacific Grove, California. The town values these butterflies so much that there is a $500 fine or a six-month sentence for killing a monarch.

Butterflies can distinguish types of sugars that humans cannot taste at all. Monarchs, for example, are

butterfly

1,200 times more sensitive to sweets than are humans.

Monarch butterflies in the larval stage feed only on the milkweed plant, giving them a bitter taste, which is an adaptation that discourages predators. Viceroy butterflies, which birds find extremely palatable, mimic the color pattern of monarchs to avoid predation.

The word *butterfly* is not a rearrangement of the words *flutter-by,* as often believed. The *Oxford English Dictionary* says that the word has its origins in the Old Dutch word *boterschijte,* referring to the yellow excretion of these insects, or that it is from the yellow cabbage butterfly that is first seen in early spring or the butter season.

Camel

Camels have a groove running from each nostril to their cleft upper lip so that any moisture from the nostrils can be caught in the mouth.

Camels are born without humps. Humps store fat to provide energy and moisture reserves for the camel.

After prolonged periods of not drinking, a camel can drink up to 20 gallons of water.

Over a four-day period a camel is able to carry 300–500 pounds at a rate of about 35 miles per day.

The characteristic rolling gait of the camel is accomplished by bringing up both legs on the same side simultaneously.

A camel is a swift animal and runs as fast as a quarter horse over a short distance. The word *dromedary* is derived from the Greek word *dromas,* which means "running."

The camel originated in North America, where it has subsequently become extinct. The living American camels, the guanaco and the vicuña, inhabit South America, while the large camels of the Old

World inhabit Asia, the Middle East, and North Africa.

Camels are unique among the mammals because their red corpuscles (blood cells) are oval shaped rather than round. This adaptation facilitates the retention of oxygen.

A camel can lose as much as 30% of its weight by dehydration without suffering adverse effects (humans will approach death if they suffer a 12% loss). They can also drink as much as 36 gallons of water in a single day.

In 1856 Jefferson Davis, then U.S. Secretary of War, imported 79 dromedary camels to the United States to carry military supplies across the deserts of the southwest. During the Civil War the camels were sold to miners and circuses and some were let wild in the desert. The camels thrived in the American deserts but were killed off by Indians and cattlemen.

Canary

Popular domestic cage birds, canaries were originally imported from the Canary Islands.

The best singing canaries in the world are bred in the Hartz Mountains of Germany.

Only male canaries are accomplished singers.

Capybara

The South American capybara is the world's largest

rodent. Capybaras can measure as long as 4 feet and weigh 175 pounds.

Highly aquatic, capybaras have webbed feet and swim with only their eyes, ears, and nostrils projecting above the surface of the water.

Capybaras frequently stand belly deep in water, feeding on various aquatic plants.

The thick, fatty skin of the capybara provides a grease that is used in the pharmaceutical trade.

Venezuelan Catholics are allowed to eat capybara meat on Church-designated meatless days, because the Church considers the rodent an amphibious animal.

Cardinal

These bright red birds were given their name by ornithologists because their color reminded them of the bright red robes worn by Roman Catholic cardinals.

Cassowary

The three species of cassowary, an ostrichlike bird, are found only in New Guinea, Australia, and adjacent islands. The largest species, the double-wattled cassowary of Australia and New Guinea, can be 6 feet tall and weigh over 100 pounds.

Though little is known about the natural behavior of the cassowary, they have been seen bathing and appear to be good swimmers.

The cassowary's wing feathers are unusual in that they consist of quills alone, which appear more like coarse hair. The advantage of these feathers is that they are not easily damaged by the dense vegetation in which the cassowary lives.

Cassowaries carry a flattened, horny ornament known as a casque on top of their heads. These casques, which can measure up to six inches high, function to protect the head from vegetation by parting the vegetation as the cassowary pushes through the forest in search of food.

Cassowaries have long claws and a daggerlike spike on each of their powerful legs, which they use in defending themselves. In New Guinea cassowaries have been known to kill people.

New Guinea tribesmen sometimes include cassowary chicks as part of the bride price paid for a wife.

Female cassowaries are generally larger than males.

Cat

Cats were probably domesticated first in Egypt around 5,000 years ago.

A cat's eyes do not glow in the dark but are very efficient in reflecting available light. This makes the eyes appear to glow.

The only domestic animal not mentioned in the King James version of the Bible is the cat.

Of the total canned fish consumed in the United States, cats eat one third.

Caterpillar

A caterpillar has 4,000 muscles, as compared to 792 muscles for a human being.

The average leaf-eating caterpillar can eat the equivalent of a square foot of leaves in a 24-hour period.

The Mexican jumping bean is actually the caterpillar or larvae of the small bean moth or *Carpocapsa saltitans* residing in the hardened bean shell. The caterpillar eats the insides of the shell, then forms the pupa, and in this state twists and turns, causing the shell to "jump."

Catfish

Catfish have been known to live as long as 60 years.

The catlike "whiskers" of the catfish, called barbels, contain sensitive nerve cells that help the catfish catch food.

Semi-saltwater European catfish grow as large as 10 feet.

Africa's Clariid catfish can live out of the water for several days. They have special organs that enable them to get sufficient oxygen. In droughts this capability enables them to seek new ponds or rivers by moving over land.

The tropical African catfish of the Mochocidae family swim on their backs in order to feed on the algae growing on the bottoms of lily pads.

Centipede

Despite their name, centipedes can have 13–95 pairs of legs.

Centipedes are poisonous.

Chameleon

Chameleons do not exist in North America. The lizard that most Americans refer to as a chameleon is the anole lizard.

The true chameleons of the Old World have long sticky tongues measuring up to 13 inches that they use to capture the insects on which they feed.

The eyes of a chameleon can move independently of one another.

Chameleons are generally arboreal and possess prehensile tails.

Cheetah

The cheetah is the fastest animal over short distances. Its top speed is 65–70 mph and it can reach 45 mph in two seconds. At full speed a cheetah runs in 23-foot strides, because its body is extremely flexible and it is able to disengage its limb bones from its shoulder girdle.

Cheetahs are easily tamed and have been used as hunting animals for thousands of years.

Cheetahs have many doglike characteristics including hard foot pads and claws that cannot be completely retracted.

Chickadee

The chickadee, or North American titmouse, can throw its voice to trick predators.

Chicken

The domestic fowl is probably the most numerous species of bird in the world.

The average hen produces about 227 eggs per year.

The progenitor of the domestic fowl is the Indian jungle fowl.

Chickens feed primarily on grains but also eat insects and small animals when the opportunity arises.

Chimpanzee

Within the genus *Pan,* there are two species, including the diminutive pygmy chimpanzee *Pan paniscus.*

In general, a chimpanzee's arm span is 50% greater than the animal's height.

Chimpanzees are gregarious, living in groups of 2–20. The composition of these groups is fluid, and members come and go from the group at will.

Chimps are normally "knuckle walkers," but they can and frequently do walk upright with their toes turned outward.

Chimps build nests in trees each night for sleeping.

The early development of a chimpanzee infant is as fast as or faster than that of a human infant.

Chimpanzees normally feed on fruits, leaves, and other vegetable material but are known to attack and kill animals as large as young bushbucks in order to eat meat. Chimps also collect and eat termites by stripping the leaves off twigs, inserting the twigs into the termites' nest, and then picking the termites off of the twigs with their lips.

Chimpanzees can live over 40 years in the wild and have lived over 50 years in captivity. One captive female gave birth to her seventh infant at just under 30 years of age.

Despite their jovial nature as youngsters, adult chimpanzees in captivity are large and extremely powerful, and are dangerous because they can be unpredictable. This, however, is not true of the pygmy chimpanzee, which remains relatively docile throughout its lifetime.

Chipmunk

According to the *Audubon Society Field Guide to North American Mammals,* a yellow pine chipmunk burrow was found to contain 67,970 food items stored for the winter.

Civet cat

Musk for perfumes is obtained from African civet cats. The word *civet* is derived from the Arabic word for perfume. Civet cats use this musk to mark their territorial boundaries.

Clam

The largest giant clam on record weighed 579.5 pounds and was found on the Great Barrier Reef of Australia. The shell of this huge clam is on display at the American Museum of Natural History in New York.

Cobra

The king cobra is the world's longest poisonous snake. The record holder was 18 feet 9 inches long. Captured in Malaysia, this cobra was later displayed at the London Zoo until the outbreak of World War II when the British government ordered all venomous snakes to be killed as a safety precaution.

King cobras primarily feed on other snakes. Their scientific name actually means "snake eater."

King cobras produce such large amounts of toxic venom that they have been known to kill full-grown elephants by biting them in the soft skin at the tip of the trunk or around the edges of the toenails.

The spitting cobras of Africa and Asia can spit venom as far as ten feet as a defensive measure. The venom will cause conjunctivitis and possibly blindness if not washed out immediately. As a safety precaution zoo keepers wear goggles when working around these snakes.

Spitting cobras are able to "spit" their venom because the holes in their small, permanent erect fangs are placed at the front of the fang rather than at the bottom.

Cockroach

Much to humankind's dismay, cockroaches have an amazing ability to survive. They have been on earth about 360 times longer than humans and yet have remained unchanged. If a cockroach's head is removed, the insect will continue to survive for several weeks before finally starving to death. They can be frozen, subsequently thawed, and will walk off. They can survive 126 G's (astronauts black out at 12) and can endure 100 times as much radiation as humans can.

The "TV roach" (Supella supellectilium) was originally an inhabitant of East Africa but is now found in many areas of the world because it has adapted to life in the backs of TV sets. This roach eats glue, insulation, and other components, never leaving the set.

Most adult cockroaches can fly. They are winged insects belonging to the order Orthoptera, which means "straight wings."

The cockroach is a living fossil. It dates back almost 300 million years.

Cockroaches eat almost anything: paper, soap, paint, glue, fingernail clippings, and human hair.

Compared to most insects, cockroaches lay comparatively few eggs, usually between 12 and 40.

Cockroaches are used in cancer research because they develop cancerous tumors similar to the ones grown by humans.

Codfish

One of humankind's most important marine food resources, the cod normally lays between 4 million and 6 million eggs in a single spawning.

Coelacanth

Believed to have been extinct for over 60 million years, a living coelacanth was caught in a trawler's net off South Africa in December 1938.

A lobed finned fish with fins attached to fleshy appendages, coelacanths are in the same group of fish from which evolved the amphibians more than 30 million years ago.

Collie

The collie received its name from Scottish breeders because the dogs were originally "coalie black."

Colugo

The colugo, or "flying lemur," of Southeast Asia is not a lemur but a primitive gliding mammal not closely related to any other living species.

Colugos are the only members of the order Dermoptera, meaning "skin wing."

Condor

Condors are an ancient type of New World vulture.

The population of the California condor, one of the world's rarest birds, today stands at around 30 individuals living in a reserve in Ventura County, California. The California condor has an extremely low reproductive rate, laying only one egg every other year.

Condors are virtually unchanged since the Pleistocene epoch over one million years ago.

Contrary to popular belief, the condor cannot fly off with a calf or lamb in its feet. A condor's feet are weak and not equipped for carrying off prey.

Coral

Coral is an animal, not a mineral or vegetable. It is in the Anthozoa group of the phylum Coelenterata and is known as the "flower animal." The term *coral,* however, generally refers to the massed skeletons of these animals.

Cow

The average dairy cow produces about 5,000 pounds of milk a year, an average of 6.6 quarts per day. A champion milker can produce about 20,000 pounds a year. An American holstein has been known to produce 339,854 pounds, or 163,245 quarts, in its lifetime.

Crab

The robber crab of the South Pacific climbs palm trees in order to feed on coconuts.

Male fiddler crabs have large "fiddle" claws that they use in combat with other males during the mating season.

The horseshoe crab is not a true crab. It belongs to the order Xiphosura and its closest living relatives are the spiders, scorpions, and other members of the class Arachnida.

Crane

The word *pedigree* owes itself to the crane. It is from the French term *pied de grue,* meaning "foot of the crane," because the three-pronged symbol indicating lines of descent vaguely resembles the crane's footprint.

The tallest bird found in North America is the whooping crane, of which only about 50 individuals survive.

Crayfish

The crayfish has an eye in its tail in addition to the two eyes on its head. The eye is called an ocellus, or pigment cup, and is retractable.

Cricket

The name *cricket* comes from the Old French verb *craquer,* which means to "creak" or "click."

Cricket chirping is done only by males and its purpose is to attract females.

A cricket's sense of hearing is located in oval slits on its forelegs.

Asian people keep crickets in small bamboo cages because they enjoy their singing. These captive crickets feed on rice, vegetables, and mosquitoes.

Crocodile

By constantly shifting from sun to shade or by basking partially submerged, the Nile crocodile keeps its body temperature within a 6° F range of its normal 78° F. By so doing, the Nile crocodile comes closer than any other reptile to the efficient regulation of body heat.

The hydrochloric acid found in a crocodile's digestive tract is so strong that it has been known to dissolve metal objects swallowed by the reptiles.

More people are killed in Africa by crocodiles than by any other animal.

Crocodiles can remain submerged for up to an hour.

Crocodiles are "living fossils" and have remained unchanged since the age of dinosaurs.

Crocodiles have been known to cannibalize their own young.

Crocodiles swallow stones as ballast and as aids in digestion. The average crocodile has 5–10 pounds of small stones in its stomach.

The fourth tooth of the lower jaw of a crocodile protrudes, one means of distinguishing them from alligators.

Crow

Crows and their relatives are believed to be the most highly evolved group of birds. Without question, crows are among the most intelligent birds.

Cuckoo

Female cuckoos lay their eggs in the nests of other birds, usually those whose eggs resemble their own. Cuckoos have a short incubation period so the cuckoo chick usually hatches before the chicks of the host species, giving the cuckoo an advantage over its nest mates.

Female cuckoos often return to the nest of the surrogate mother who raised them to lay their own eggs.

Cuscus

Cuscuses are arboreal marsupials distributed on islands from New Guinea west through the Celebes.

Even when handled carefully, cuscuses usually emit a strong, musky odor.

Cuscuses are eaten by New Guinea natives. Besides humans, the primary enemy of the cuscus is the python.

Deer

Deer inhabit North America, South America, northwest Africa, Eurasia, Japan, the Philippines, and most of Indonesia. They have been introduced to New Guinea and New Zealand and reintroduced to Great Britain, where they were found until a century ago.

All deer with the exception of the musk deer and the Chinese water deer possess antlers that are grown and shed annually. Antlers are borne by the males, except in the caribou and reindeer, where both sexes are antlered.

A deer grows its first set of antlers at one to two years of age. These first antlers are often single spikes. The antlers become larger and acquire more points (called tines) as the animal matures.

The largest living deer is the moose (called the elk in Europe).

The Père David's deer of China has been extinct in the wild for over 3,000 years. Surviving Père David's deer were kept in the Chinese emperor's imperial hunting park from that time on until their existence was made known to European zoologists by a French missionary, Père Armand David, in

spotted fallow deer

1869. All Père David's deer in China were exterminated during the Boxer Rebellion, and those that survive today are the descendants of 19 animals gathered together and bred by the eleventh Duke of Bedford at his estate at Woburn Abbey in England.

The extinct Irish elk *(Megaloceros),* which lived in Ireland during the Pleistocene, had antlers that measured over 12 feet from tip to tip.

Although the musk deer of Asia lack antlers, the canine teeth of the males are well developed, measuring about three inches in length.

In Elizabethan English the word *deer* was a general term used to refer to any animal. In modern German this is still true: the word *Tier* refers to any animal; for example, *Fingertier* means the aye-aye.

The word *buck,* meaning a dollar, comes from the days when deerskins were worth one dollar and were a medium of exchange.

Lapps ranch reindeer for their meat and milk. Reindeer milk has a fat content that is four to five times higher than cow's milk.

Deer antlers that have been shed are rarely found, because rodents, foxes, and even deer gnaw them for their calcium content.

In Hindi the word *barahsinga* means "twelve-horned." It is applied to the Barahsinga deer because mature specimens normally have 12 tines on their antlers.

Deer Bot-fly

The deer bot-fly is the fastest flying insect, reaching a speed of 36 mph.

Dingo

The dingo is a semidomestic dog inhabiting Australia. Dingoes were brought to Australia around 3,500 years ago by aborigines who may have originated in Polynesia.

Dinosaur

Not all dinosaurs were mammoth. Some were as small as domestic chickens.

The vast majority of dinosaurs were vegetarians, with only a small percentage of dinosaur species adapted for eating meat. In this regard, dinosaur populations were very much like modern mammalian populations, leading scientists to speculate that dinosaurs were warm-blooded.

A pterosaur discovered in Texas is possibly the largest flying animal that every lived. Living over 60 million years ago, this species had a wingspan of over 27 feet.

The brain of a stegosaurus was smaller than a nerve knot on its spine.

Dodo

The extinct dodo of Mauritius received its unusual name from the Portuguese word *doudo,* meaning "simpleton," a reference to its tameness and odd looks.

Before the seventeenth century three species of dodo were living on the islands of Mauritius, Réunion, and Rodriguez in the Indian Ocean.

Dodoes were about the size of a turkey and were closely related to the pigeons.

Dodoes fed on fallen fruits and leaves and may have taken some animal food, such as crabs.

The males of the Rodriguez solitaire, a type of

dodo, developed round, bony outgrowths toward the ends of their small rudimentary wings.

Dodoes mated for life.

Dog

There are about 35 living species of canids. They are worldwide in their distribution.

Among the canids, the fennec fox is the smallest and the timber wolf is the largest.

The Falkland Island wolf *(Dusicyon australis)* was exterminated during this century, mostly by the actions of fur traders and sheep ranchers.

The dog has been a domesticated animal since prehistoric times. All domestic dogs belong to the single species *Canis familiaris.*

A dog's sweat glands are located in the nose and on the bottom of its feet. Dogs regulate their temperature by panting.

The heaviest domestic dog on record was a 295-pound St. Bernard.

The poodle was originally bred as a retriever.

The miniature schnauzer was bred in Germany as a rat terrier. *Schnauzer* is the German word for "growler."

Dolphin (Mammal)

Dolphins are actually small, toothed whales.

Dolphins have been known to kill sharks by ramming them with their snouts.

There are about a dozen species of dolphin that live permanently in fresh water in large rivers such as the Ganges and Amazon.

Dolphins have extremely high IQs, though it is at present impossible for humans to determine how intelligent dolphins actually are.

Like all whales, dolphins lack vocal cords and communicate by producing high frequency clicks and chirps through their blowholes.

Dolphins use echolocation in order to communicate and to search for food.

Dolphin Fish

The dolphin fish, which can measure over six feet in length, is sold in restaurants as mahimahi.

The sex of a dolphin fish can be determined by examining the shape of its forehead. Males have high domed foreheads and females have gradually sloping foreheads.

Dormouse

The dormouse's name is derived from the Latin word for sleep, since dormice spend six to seven months of the year sleeping.

Romans used to fatten dormice by feeding them large quantities of oilseed before eating them.

Dove

The word *dove* is used to refer to the smaller members of the pigeon family; otherwise it has no specific scientific meaning.

When drinking, doves and pigeons suck up water. Most other birds have to throw their heads backward to swallow water.

Dragonfly

The dragonfly is not a true fly. True flies have one pair of wings, while dragonflies have two.

Millions of years ago a type of dragonfly existed with a wingspan of four feet; it was the largest insect ever to have inhabited Earth.

The dragonfly lives two years as an aquatic grub but only about 12 days as an adult dragonfly.

Dragonfly eyes have 28,000 fixed lenses and can see 360° around them.

Dragonflies are believed to be one of the oldest insect forms and the first animals to attain flight.

Dragonflies cannot walk, and use their legs only for perching.

The largest dragonfly today, *Megaloprepus caerulatus,* of Central and South America, has a wingspan of about seven inches.

Drumfish

The drumfish, a freshwater species, has teeth in its

throat, which it uses to grind down the shells of mollusks, on which it feeds.

Duck

Almost all ducks line their nests with down plucked by the female from her own breast.

In Iceland huge colonies of eider ducks *(Somateria mollissima)* are protected and farmed for their down. Each time the female lines the nest, the lining is harvested, though not to the point where the female would abandon the nest.

Water runs off a duck's back because its feathers are naturally well oiled.

Ducks are comfortable swimming in near-freezing water, yet their body temperatures are higher than humans'. Their feathers provide excellent insulation and keep the cold water from ever touching their bodies.

Most domestic ducks are descendants of wild mallards.

Some ducks can fly at speeds approaching 70 mph.

Dung Beetle

The absence of dung beetles in Australia once created an agricultural problem. Thirty million cattle rendered six million acres of land unusable because of the omnipresence of cow dung. Imported dung beetles saved the situation.

Eagle

Bald eagles, the national symbol of the United States, primarily feed on fish. They mate for life and may pass on their massive nests to subsequent generations.

Many eagles build huge nests; the nest of one pair of bald eagles measured in Ohio was 8½ feet in width and 12 feet in depth and weighed 2 tons.

The most powerful eagles, like the Philippine eagle, can kill animals approximately 5 times their own size.

Eagles are large members of the hawk family.

The Indian black eagle is specialized to feed on birds' nestlings and eggs.

Earthworms

Earthworms are hermaphroditic, each individual possessing a set of male and female genitalia.

Giant earthworms found in Australian rain forests can be as long as 11 feet and stretch to 21 feet. Their eggs are about the size of olives.

Earthworms eat and process their own weight in food every day. A ton of earthworms can turn a ton of organic garbage into useful soil.

Eel

The freshwater eels of the United States and western Europe are all hatched in the Sargasso Sea in the Atlantic Ocean. Adults migrate there from the lakes and rivers they inhabit. Newborn eels return to the same areas from which their parents came, usually taking one to three years to make the journey.

The electric eels of South America use the electricity they produce with specialized muscles to locate and stun their prey.

Elephant

Elephants are the largest terrestrial mammals living today. The largest known elephant was a male shot in Angola in 1955, which is now exhibited at the National Museum of Natural History in Washington, D.C. This massive bull stands 13 feet 2 inches at the shoulder (the average size for an African bull is 10.5 feet), measures 33 feet 2 inches from the tip of its trunk to the tip of its tail, and was estimated to have weighed 24,000 pounds in life.

The skeleton of an elephant is necessarily massive, comprising 12–15% of the animal's total body weight.

African elephant

The trunk of the elephant is a remarkable organ, having evolved from the nose and upper lip. The trunk consists of 40,000 muscles and tendons, which make it strong, flexible, and extremely agile. The trunk of the Asian elephant ends in one fingerlike projection at the tip, while that of the African species has two.

The ears of the African elephant can measure up to 30 square feet in surface area. Each ear is infused with numerous blood vessels and acts as a cooling organ when flapped. Each ear may weigh as much as 110 pounds.

The term *pachyderm* is often used as a synonym for elephant, but the term actually refers to any thick-skinned hoofed animal such as the rhinoceros and hippopotamus.

Elephants have the longest gestation period of any mammal, up to 24 months.

Like humans, elephants have a life expectancy of between 60 and 80 years.

The tusks, which are found in both sexes of the African elephant and in males of the Asian species, are specialized incisor teeth that continue to grow throughout the animal's life. The longest tusks on record were a pair of African elephant tusks that measured 11 feet 5 inches and 11 feet respectively and had a combined weight of 293 pounds.

Elephants consume large quantities of grass, vines, tree shoots, and fruit. An elephant can eat up to 500 pounds of forage in a day.

The word *elephant* is derived from the Greek word *elephas,* meaning "ivory." The Swahili word for elephant is *tembo.*

The largest African elephant ever held in captivity was the famous Jumbo, exhibited in North America by P. T. Barnum after he purchased Jumbo from the London Zoo. Jumbo stood 11 feet at the shoulder and weighed 6.5 tons. Jumbo was killed at St. Thomas, Ontario, on September 15, 1885, when he was struck by a locomotive. The force of the impact was so great that the locomotive and two cars were derailed and the engineer was killed. Jumbo's skeleton is now on display in the American Museum of Natural History in New York.

Although thick, an elephant's skin is extremely sensitive and must be carefully looked after in captivity.

At certain times of the year bull Asian elephants go into a sexual heat called musth, which is characterized by an oily discharge produced by a gland located between the eye and ear. At these times the bulls are completely uncontrollable and extremely dangerous. This is the main reason why very few bull elephants are kept in captivity.

The molar teeth of elephants are huge (six to seven inches). At any given time only one tooth is in use in the upper and lower jaws of the elephant's mouth. As the tooth is worn away, it is pushed forward and gradually replaced by a new tooth from behind.

White elephants do exist; some are true albinos, some are very pale in color. The earliest recorded white elephant was in 1636; it was one of 6,000 owned by the Siamese royal palace. This elephant was considered so valuable that it was cared for by one of the Princes of the Blood. Buffon reported in his *Natural History* that in Siam, the white elephant was considered so special that the only human being before whom it must bow the knee was the emperor himself and the emperor always returned the salute.

P. T. Barnum once featured an albino elephant in

his show, but it proved to be a modest attraction and was expensive to maintain and hard to get rid of; hence the origin of the expression "white elephant."

Elephant Bird

The elephant bird (Aepyornis) of Madagascar was the world's heaviest bird, weighing as much as 1,000 pounds.

Some scholars believe that the fabulous rocs described in the Travels of Marco Polo were actually secondhand accounts of the elephant bird.

Early European travelers in Madagascar described how the Malagasy natives used the eggshells of Aepyornis for water flasks. Each egg is capable of holding up to two gallons of liquid.

Though it is not known precisely when the last elephant bird died, it is believed that hunting by the early inhabitants of Madagascar contributed to their demise. One species may have survived as late as 1649.

The largest species of elephant bird was Aepyornis titan. Its eggs are the largest single cell known to science.

Eulachon

The eulachon (Thaleichthys pacificus), a type of smelt, has a skin that is so oily that American Indians use it as a candle.

European Swift

The European swift can dive at speeds approaching 200 mph.

European Weatherfish

European weatherfish *(Misgurnus fossilis)* are very sensitive to changes in barometric pressure and get restless during a decrease in pressure, usually indicating a storm.

Fairy Tern

Fairy terns do not construct nests; they lay their single egg directly onto the forked branch of a tree.

Female fairy terns remain with the egg continuously because that is the only way to prevent it from rolling off the limb as it sways in the wind.

Fairy tern chicks are precocious and perch unaided on the limb from the moment of hatching.

Falcon

Some species of falcons have serrated upper mandibles, which assist them in killing prey.

The forest falcons *(Micrastur)* have large eyes and specialized ears that aid them while hunting in almost total darkness. These adaptations are shared by the other group of nocturnal avian predators, the owls. Other falcon species are active during the day.

The largest falcons weigh 3–4.5 pounds, while the smallest falconets weigh only 1.5 ounces.

When diving for prey, the peregrine falcon can attain speeds of 175 mph.

Firefly

The firefly is not a fly but a beetle.

In general, female fireflies do not fly. They lie on the ground and attract males with their bright glow. The few female fireflies that do fly do not normally glow as brightly as males of their species.

The female firefly of the genus *Photouris* mimics the signals of the genus *Photinus,* and when the male *Photinus* gets close, the female *Photouris* devours him.

When English explorers Sir James Cavendish and Sir Robert Dudley first landed in the West Indies, they saw what appeared to be Spanish settlers holding matches and moving toward them. They immediately returned to their ships. What they actually saw were vast numbers of fireflies brightly glowing.

Fireflies live only for a few days. Their glowing lights are signals to attract the opposite sex and to ensure correct pairing of each species.

Fireflies produce a cold light; 98% of the energy is used for making light, and only 2% produces heat.

Fish

Fish outnumber all other vertebrate animals, with approximately 30,000 living species. Of these about 2,300 are freshwater species.

Lacking eyelids, fish sleep with their eyes open, apparently undisturbed by visual images to the brain.

Black sea bass *(Centropristes striatus)* are predominantly egg-producing females when young, but at around five years of age most switch sex and become functional males.

Fish are sensitive to low-frequency vibrations and pressure waves produced in the water by solid objects. This is why fish do not swim into the walls of aquariums.

The world's earliest known vertebrate animal was the jawless fish *Agnathans*, which lived some 480,000,000 years ago.

The largest species of fish is the plankton-eating whale shark *(Rhincodon typus)*, inhabiting the warmer waters of the Atlantic, Pacific, and Indian oceans. The largest whale shark ever recorded was a 59-foot specimen killed in the Gulf of Siam in 1919.

The skin of a whale shark is four to six inches thick.

Drumfish produce loud noises by beating stringlike muscles against their air bladders, which act as resonators.

At the time of hatching, a fish has its full quota of scales; as the fish grows, the scales grow proportionally.

Female sea horses deposit their eggs in a specialized "brood pouch" on the male's abdomen, where they are fertilized, developed, and born.

Sea horses are the slowest-swimming fish known.

The fastest-swimming fish are the sailfish and

swordfish, which are able to reach speeds in excess of 60 mph.

The white sturgeon, the largest North American fish, may spend as long as 20 years maturing in the ocean before migrating up large rivers to spawn. Adult white sturgeon may be as long as 12.5 feet and weigh over 1,200 pounds.

The eggs of the freshwater gar are poisonous if eaten.

The four-eyed fish of Central America (Anableps dowei) lives at the surface of fresh water rivers and has eyes adapted for seeing both above and below the surface simultaneously. The upper part of the eye lens is flattened for seeing through air, while the lower part of the lens is rounded for sight in water—natural bifocals.

The stonefish of the tropical Indo-Pacific region possess the most toxic venom of any known fish.

Fish continue to grow throughout their lifetimes.

The largest freshwater fish in the world is the arapaima (Arapaima gigas), inhabiting the major rivers of South America. Large specimens measure up to eight feet in length.

Pandaka pygmaea, a type of goby inhabiting streams and lakes in the Philippines, is the world's smallest known vertebrate, measuring no more than one-half inch as an adult.

The snail darter, a small fish found in the Little Tennessee River, delayed construction of the Tellico Dam project in the late 1970s because it was an endangered species. After the project was completed, however, snail darters were found also to inhabit rivers in Alabama and Georgia. There are now efforts to reclassify the snail darter as a threatened rather than an endangered species.

Flamingo

Flamingos belong to the family Phoenicopteridae, one of the oldest bird families still living.

A flamingo's bill is highly specialized for filter-feeding with the lower mandible large and trough-like and the upper mandible thin and lid-like. Flamingos feed on small invertebrates, crustaceans, and algae living in brackish water.

There are six species of flamingo inhabiting the tropical Americas, India, and Africa. They range in size from three to six feet.

Flamingos are among the most social of all birds and will seldom breed in flocks of less than 20 birds.

Flea

A flea can jump 7 inches into the air or broadjump 13 inches—300 times its own length. Jumps are powered by resilian, a mass of rubberlike protein that decompresses very rapidly, pushing the flea into a cartwheel spin.

Flounder

The plaice, a type of flounder found in Europe, can camouflage itself so well that when placed on a checkerboard pattern, it will assume the pattern itself.

Fly

Flies eat only liquid foods like nectar, honey, sweat, and blood.

In general, the smaller the fly, the faster its wings beat. The midge beats its wings about 1,000 times a second, and the housefly beats its wings about 200 times a second.

The Japanese ichneumon fly has a tail that is almost ten times the length of its body. The tail consists mostly of an ovipositor used for laying its eggs in the ground.

Flying Cuttlefish

The flying cuttlefish has the ability to move by jet propulsion. It pushes water through its system with such force that it can move at speeds of 20 nautical mph and jump 15 feet out of the water.

Flying Fish

Flying fish don't actually fly but they can glide over

the surface of the water for distances of over 500 feet.

Fox

Foxes are in the family Canidae, as are wolves, dingoes, and the domestic dog.

In French the fox is called *renard* from the poem of the Middle Ages, *Reynard the Fox.*

In English a male fox is called a reynard and the female is called a vixen.

Some Japanese department stores have fox shrines on top of their roofs to propitiate the animal, to which the Japanese ascribe magical powers.

Despite its name, the crab-eating fox feeds primarily on rodents, grasshoppers, crickets, lizards, and snakes.

A fox has excellent senses of hearing, sight, and smell, which is why they are such successful and adaptable predators.

Frigate Bird

The frigate bird has the lightest skeleton relative to its wingspan of any bird. A frigate bird with a seven-foot wingspan has a skeleton weighing about four ounces, less than the weight of its feathers.

Frog

There are about 1,800 species of frogs and toads, making them the largest order of amphibians.

The goliath frog of West Africa is the largest of the approximately 250 species of frogs. With legs outstretched, the goliath frog can measure up to 32 inches and weighs as much as 7 pounds. They inhabit fast-running water around waterfalls and rapids.

Goliath frogs lack vocal cords and do not croak.

The hairy frog of Africa has many fine filaments of skin infused with blood vessels on its sides and back legs. The "hairs," which number about 2,500, more than double the area of the frog's skin surface, are an adaptation for greatly increasing the amount

meadow frog

of oxygen that the frog can absorb from the water in which it lives. Only males possess these hairs, because the females normally live on land and only join the males in the water during the breeding season.

Giant Sable Antelope

The giant sable antelope of Angola has the longest horns of any antelope. The record length of the horns of a giant sable is 65 inches.

Gibbon

The gibbons of Asia are the smallest and most arboreal of the apes. They are also considered to be the most agile of all mammals in the trees.

The generic name of the six gibbon species is *Hylobates,* which means "dweller in the trees."

Gibbons progress by swinging arm over arm through the branches of trees, a method of locomotion known as brachiation. To help them brachiate, gibbons have elongated arms and hands, and their thumbs are reduced in size so as not to get in the way when they are moving rapidly.

Gibbons mate for life and are highly territorial. Each pair of gibbons establishes a territory, where it lives with one or more of its immature offspring. Gibbons will not tolerate other gibbons within their territorial boundaries and will vigorously attack intruders. A territory's boundaries are proclaimed in loud vocal chorus sung by both males and females in the early morning.

Gibbons are so quick that they have been observed catching flying birds in midair and completing their leap to the next branch without a break in rhythm.

Gibbons feed on a diet composed of fruits, leaves, buds, insects, birds, and eggs.

Gila Monster

The gila monster of the southwestern United States and its close relative, the beaded lizard of Mexico, are the only known poisonous lizards in the world.

These lizards do not possess hypodermic fangs but inject venom by means of grooved teeth in the lower jaw. This system means that they must hold on and chew to inject their venom.

The gila monster is so named because it was once very common along the Gila River basin.

Although poisonous, gila monsters and beaded lizards are placid and slow-moving by nature, and there is no authentic record of a human dying as the result of the bite of one of these lizards.

Giraffe

The family Giraffidae comprises the giraffe and the forest-dwelling okapi, both natives of Africa.

The horns of the giraffe are unlike those of any other mammal. They are present at birth as cartilaginous knobs, then turn to bone shortly after birth and grow slowly thereafter throughout life. Horns are present in both sexes but are smaller on females. Horns are covered with hair and skin and there is an additional third horn in the center of the forehead.

Giraffes are the tallest of all living mammals; bulls may attain a height of 19 feet and a weight of over 2 tons.

Despite the great length of a giraffe's neck, it is composed of only seven vertebrae, the same number as in humans and most other mammals.

Giraffes are thought by some to be mute, but in fact they are capable of uttering low moans and bleats.

The giraffe may have the keenest vision of any African game animal, and its great height gives the giraffe the greatest range of vision of any terrestrial animal.

Giraffes normally live in groups of 15–20 individuals led by a mature bull.

When walking, giraffes pace like camels, simultaneously moving the fore and hind leg on the same side of the body.

In defending themselves, giraffes kick with their forefeet. When fighting for the possession of cows, bulls assail one another with blows delivered with the head and neck.

giraffe

Giraffes are browsers and cud-chewers. As an adaptation for their feeding pattern the tongue of the giraffe measures up to 18 inches in length. If water is available, giraffes will drink occasionally, but they are able to go without water for many weeks.

The specific name of the giraffe is *camelopardalis,* which is Latin, meaning "camel marked like a leopard."

Giraffes have extremely high blood pressure because of their long necks; the pressure is so high that giraffes have special valves in their aortas to regulate blood flow to the brain.

Female giraffes give birth in a standing position and newborn calves drop six feet to the ground.

Before the great Ice Age, giraffes roamed in central Asia and Europe as well as in Africa.

Giraffes do not swim, so large rivers are barriers that limit their distribution.

The word giraffe is derived from the Arabic *zarafa,* meaning "swift creature"; giraffes can run at speeds of up to 30 mph.

In captivity, giraffes can be kept in simple, barless enclosures because they are unwilling to step down more than about 18 inches.

Globefish

The globefish can greatly increase its size by rapidly swallowing water in order to defend itself from predators.

Glowworm

The glowworms living in the caves at Waitomo on the north island of New Zealand hang gummy threads from their bodies. The glow from the threads attracts insects, which get stuck and are subsequently eaten by the glowworms. These millions of glowworms and their threads create the effect of a beautiful glowing curtain.

Gnat

The common gnat *(Culex pipiens)* is actually a mosquito. In fact, the word *gnat* is Old English for "mosquito." The gnat is capable of biting but usually reserves its bite for birds and not humans.

Gnatcatcher

The blue-gray gnatcatcher, a bird found in tropical areas of the southern United States and northern Mexico, uses spider webs to help bind the lichens with which it builds its nest.

Gnu

The two species of gnu are the white-bearded gnu, one of Africa's most numerous game animals, and the white-tailed gnu of South Africa, which was almost exterminated at the end of the last century but whose numbers are now stable as a result of propagation in captivity.

When disturbed, gnus behave oddly, displaying curious dances during which they prance, paw the ground, dig their horns into the earth, and thrash their tails before running away. This behavior led to their being named *wildebeest* in Afrikaans.

In captivity, gnus are unpredictable and occasionally dangerous. One white-tailed gnu killed its keeper in a zoo.

The name *gnu* is a Hottentot word that mimics the animal's call.

Goat

Cashmere wool comes from a goat. It takes the wool from about 40 goats to produce an overcoat.

Goat's milk is more easily digested than cow's milk, does not require homogenization, and is often recommended for infants and people who have difficulty digesting cow's milk.

Goldfish

Goldfish are a species of carp.

Goldfish have been domesticated for thousands of years in Asia.

In its natural state the goldfish is usually greenish brown or gray.

In captivity, goldfish have been known to live over 80 years.

Goliath Beetle

The goliath beetle, the world's heaviest insect, nearly ¼ pound (3½"–4"), has been known to peel a banana while in captivity, as did "Buster the Bug," a goliath beetle mysteriously found in a coffee can left on the steps of the American Museum of Natural History in New York. The immigration department wanted to deport the beetle, but the chief museum entomologist, Dr. John C. Pallister, persuaded the officials to let the beetle remain at the museum.

Because the goliath beetle makes a loud whirring sound, youngsters in Africa often tie them to strings and fly them in circles around their heads.

Goose

Migrating geese and other birds fly in the classic V formation for a very good reason. Except for the leading goose it makes the flying easier for all the other birds. The air currents left by each bird make it easier for the bird behind it to fly. Because the bird in front has to work harder, they usually take turns in the lead position.

During migrations geese have been observed flying as high as 29,500 feet over the Himalayan Mountains in temperatures around −12°F.

The magpie goose *(Anseranas semipalmata)* of northern Australia has baffled ornithologists for years because of its unusual characteristics. Magpie geese have partially webbed feet and relatively long legs, perch and roost in trees, and are able to fly during the molting period.

The state bird of Hawaii is the nene goose, an upland species that was nearly exterminated in the early part of this century. It is now off the endangered list because it was bred in captivity and eventually returned to its former range. Like the magpie goose, the nene also has partially webbed feet.

Gorilla

The gorilla is the largest living primate. Of the three subspecies recognized by zoologists, the largest is the Eastern lowland gorilla, *Gorilla gorilla graueri*, which can stand as tall as 6 feet and weigh up to 450 pounds (females are normally one half the weight of males).

The gorilla was the last of the great apes to be described by Western science. The Western lowland gorilla was first described by Savage and Wyman in 1847; the Eastern races were not described until 1902.

Despite their large size as adults, infant gorillas average between two to four pounds at birth. The largest gorilla ever born in captivity is the male Jim, born at the San Diego Wild Animal Park in 1973 and now living at the San Diego Zoo. At birth Jim weighed 6¼ pounds.

gorilla

Despite their fearsome reputations, gorillas are gentle, highly intelligent, and are vegetarians. Gorillas are big primarily because of the types of food they eat. Their tremendous size is an adaptation that aids them in collecting and preparing the tough, fibrous vegetable material they consume. An adult gorilla has no problem breaking down a bamboo that may be 5 inches in diameter and stand over 30 feet tall or tearing apart a palm in order to consume the contents of its pithy trunk.

A gorilla's skull is a marvel of evolution, well adapted for the consumption of coarse vegetable material. The mandible is thick, containing strong teeth that, although larger, are virtually identical to those of humans. In order to chew the plants on which they feed, gorillas require massive jaw muscles connecting the mandible with the top of the skull. These muscles are so large that there is not sufficient room for their attachment on the gorilla's brain case. As a result, as a gorilla matures, the actions of the muscles gradually mold the bones of the skull into a high sagittal crest that provides additional surface area for the attachment of muscle and gives adults the characteristic crests on top of their heads. The large brow ridges of the gorilla are infused with sinus cavities and act as shock absorbers for the massive jaws.

The gorilla's tremendous size makes it virtually invulnerable to predators; their only enemies as adults are leopards and humans.

Gorillas do not swim, so large rivers are barriers that limit their distribution. In 1951 an adult male gorilla drowned when he accidentally fell into the water-filled moat at the front of his enclosure.

Gorillas live in family groups of 5–30 individuals led by a dominant male who directs their movements and feeding patterns and acts as defender of the troop. At night gorillas construct nests in the trees or on the ground that are used for one night only.

The first gorilla ever seen alive outside of Africa was a female named Judy who was exhibited by Wombwell's Travelling Menagerie in England in 1855. During her lifetime Judy was exhibited as a chimpanzee and was only discovered to be a gorilla after her death.

Chest beating is a behavioral characteristic unique to gorillas. Chest beating can be an expression of exuberance, intimidation, or annoyance. Adult males chest-beat during intimidation displays when confronting threats to the troop.

Today the mountain gorilla *Gorilla gorilla berengei* is the most endangered species of great ape, with no more than 200–250 surviving around the Virunga Volcanoes, which form the border between Uganda, Rwanda, and Zaire. They are threatened due to human encroachment on their habitat and, increasingly, by the activities of poachers. There are less than 12,000 gorillas of all types alive today; it is almost certain that they will become extinct in the wild before the turn of the century unless more stringent conservation measures are undertaken.

Adult male gorillas are called silverbacks, because the hair on the lumbar region of their backs turns gray at the onset of sexual maturity from 10 to 12 years of age.

The oldest gorilla in captivity is 50-year-old Massa, who is still living at the Philadelphia Zoo.

In 1966 a white infant gorilla was discovered in Río Muni after his mother had been shot during a raid on a banana plantation (gorillas eat the banana trees, not the fruit). Named Copito de Nieve (Little Snowflake), he was taken to the zoo in Barcelona, Spain, where he now lives. Copito has fathered several black offspring, all of whom possess the recessive gene for white coat color.

Like humans, gorillas are born with no instinctual behaviors. Everything they must know in order to survive as adults has to be learned. This necessary learning takes place during the gorilla's prolonged period of childhood and adolescence.

Grasshopper

A grasshopper has around 900 different muscles; humans have 792.

Grasshoppers have the ability to throw their voices merely by adjusting the angle of their forewings. This device is used to confuse predators.

American Indians of the western states frequently ate grasshoppers. They usually roasted them in large numbers and ground them into meal.

Some grasshoppers can grow as long as six inches.

A grasshopper can jump over 500 times its own height.

Greyhound

The origin of the greyhound's name has nothing to

do with the color gray. According to the *Oxford English Dictionary,* the name is derived from the Old Norse word *groy,* which meant "bitch" and later came to mean "dog" in general.

Groundhog

The groundhog does not drink water. It obtains necessary moisture from the food it eats.

Groundhog is a common name often used to refer to prairie dogs, a type of ground squirrel. Prairie dogs are highly gregarious, often living in large communities. One of these communities covered a 100 by 400 mile area and contained an estimated 40 million inhabitants.

Grouse

The ruffed grouse feeds on bane berries, which are fatal to humans.

The Oregon ruffed grouse produces a loud drumlike sound that can be heard over great distances. The sound is produced when the bird strikes its wings against its chest.

During winter the scales on a ruffed grouse's toes grow out, forming functional snowshoes.

Grunion

The grunion of southern California is one of few types of fish that goes onto land to spawn. Grunion ride high tides onto the shore where the females bury their tails in the sand and deposit their eggs. Males fertilize the eggs while they are being laid, the entire process taking about one minute. Ten days later the eggs hatch and the baby grunion are washed into the ocean on another high tide.

Guinea Pig

Like humans but unlike most other animals, guinea pigs need vitamin C in their diets. Other animals produce enzymes that convert carbohydrates into vitamin C and therefore do not need the vitamin in their diets.

Guinea pigs are born with their second set of teeth in place, enabling them to eat grain within forty-eight hours after birth.

Guinea pigs were originally domesticated by South American Indians as a food item. Early Spanish explorers took these rodents back to Europe for the same reason.

Gull

Originally gulls fed on fish, shellfish, invertebrates, insects, and small animals along the seashore and inland waterways, but with the spread of humans gulls have become scavengers, living to a large extent off the wastes of humans. These habits have greatly increased the numbers of gulls and their range.

Many large species of gull have a red spot near the tip of their beak that chicks peck to stimulate the adult birds to regurgitate food.

Guppy

The guppy, a small fish popular with young tropical fish collectors, was named after Rev. R. J. Lechmere Guppy, who discovered the species on the island of Trinidad in 1866.

Hammerhead

The hammerhead, an African swamp bird, builds a huge nest that may be six feet wide and four feet high. The nest is made of twigs, woven grass, mud, and dry vegetation and is usually built in the fork of a tree. A pair of hammerheads may spend up to six months constructing their nest.

In some parts of Africa natives regard hammerheads with awe and believe that to molest a hammerhead brings bad luck.

Hamster

The word *hamster* is from the German word *hamstern,* meaning "to hoard." Hamsters have been known to hoard as much as 65 pounds of seed and grain for use during the winter.

The golden hamster has the shortest gestation period of all mammals except the marsupials, about 15 days.

Golden hamsters are descendants of a single pair of hamsters found in Jerusalem in 1930.

Hawk

With some 217 species the family Accipitridae, containing both the hawks and eagles, is the largest family of diurnal birds of prey.

Hawks are extremely important controllers of the world's rodent population.

Some hawks are highly territorial. Territories are maintained by perching in conspicuous view of neighbors, by display flights near territorial boundaries, and by vocal calls.

Hedgehog

The hedgehog rolls itself into a ball when it is threatened, causing the spines on its back to form a protective shield. Foxes, however, have been known to roll the hedgehog into the water, forcing it to unroll.

Wild hedgehogs are usually badly infested with fleas.

Hedgehogs seem to have an extremely high tolerance for toxic substances, eating poisonous animals such as bees, wasps, blister beetles, and oil beetles. They also prey on the poisonous European viper, though hedgehogs are occasionally killed themselves in attacks on vipers.

Hippopotamus

The family Hippopotamidae contains two genera:

Hippopotamus, the large hippo, which lives in river habitats in Africa, and *Choeropsis,* the pygmy hippo, which lives in dense, swampy forests in West Africa.

Male large hippos frequently fight with one another, using their large lower canines as slashing weapons. Bulls have been known to kill one another during such fights.

Hippos have an excellent sense of smell but rather poor eyesight.

The sparsely haired skin of the hippopotamus contains special pores that secrete a pinkish, oily fluid known as "blood sweat." This secretion protects the hippo's skin, allowing the animals to remain in water or in a dry atmosphere for extended periods of time.

Large hippos generally spend the day in water but emerge at night to feed on land, consuming large quantities of vegetation and often covering many miles in a night's foraging.

Baby hippos are born underwater, swim before they can walk, and nurse underwater. The young of the large hippo often climb on their mother's back and sun themselves while she is floating in the water. This behavior may afford the young some protection from crocodiles.

Hippos are hunted for their excellent flesh, abundant fat, durable hide, and for the ivory of their canine teeth.

The hippopotamus is the largest living nonruminating even-toed ungulate (hoofed) mammal. They can reach a length of 14 feet and weigh 5 tons. Pygmy hippos measure only a little over 5 feet and weigh about 525 pounds.

Hippos can run at great speeds for short distances.

The Budapest Zoo is built around a thermal spring that has proven to be an excellent breeding environment for hippos. As a result, the Budapest Zoo has supplied hippos to other zoos for many years.

The word *hippopotamus* is Greek, meaning "river horse."

Hoatzin

The hoatzin of South America is an unusual bird that has long caused confusion for taxonomists (zoologists who classify animals).

Hoatzins have weak feet and often support themselves on their breasts, which are covered with thick skin.

Soon after hatching, hoatzin chicks leave the nest and begin climbing about in the trees. While climbing, they utilize wing claws, much like those on the wings of the archaeopteryx, located at the bend of the wing. As the birds mature, these unusual claws are shed.

Hornbill

Hornbills are among the most spectacular of all birds. Some 45 species inhabit tropical Asia and Africa. They range in size from about 15 inches to almost 5 feet.

The helmeted hornbill of Asia is the only species with a solid casque (large ornamental knobs surmounting the bill) on its bill. The ivory of the helmeted hornbill has been highly valued as carving material for centuries. The first product known to have been exported from the island of Borneo was helmeted hornbill skulls for Chinese carvers.

The nesting habits of hornbills are extraordinary. Having found a suitable hole in a tree, the female lays one to five eggs, and then she and the eggs are sealed in the hole by the male, who builds a wall of mud across the entrance to the hole, leaving only a narrow slit. The female remains confined during the entire incubation period and for part of the chicks' fledgling stage, a period of up to 100 days. During this time she is entirely dependent on the male for food.

Because of their unique nesting habits hornbills are a symbol of marriage and fidelity in many parts of the world.

The large ground hornbill of Africa is the only species in which the female is not sealed into the nest during incubation. Ground hornbills have lived for over 40 years in captivity.

Horse

Horses are the only mammals that have only one functional digit. The hoof is a highly evolved and specialized middle toe.

The Mongolian wild horse was probably domesticated around 3000 B.C.

The tarpan *(Equus callabus),* now extinct, originally inhabited much of Europe and portions of Asia. This animal was an important food source for prehistoric humans and was probably domesticated very early. The scientific name of this wild horse is used for domestic horses, though domestic horses are probably a mixture of two or three wild species.

Eohippus (meaning "dawn horse"), the progenitor of modern horses, lived around 50 million years ago and was about the size of a terrier.

The horse evolved in the Americas but became extinct here about 10,000 years ago. It reappeared in the sixteenth century when Spanish conquistadores brought it from Spain. The wild horses of the American west are descendants of those horses. The only true wild horse living today is the Mongolian wild horse.

Until A.D. 420, when the stirrup was invented, horses, like camels, dropped to their knees when mounted by riders. Horseback riders mounted on the left side because early nomadic warriors carried their swords on their left sides and therefore had to mount their horses on the left side.

A pony is a small horse under 14.2 hands or about 57 inches.

Mare's milk is used to make humiss, a fermented beverage consumed in Mongolia. Some peoples drink mare's milk as is.

Large draft horses were bred originally as battle horses. They had to be large and strong enough to carry men wearing armor.

There are approximately 17,000 wild horses living in nine states in the western United States.

The oldest domestic horse lived to the age of 62.

Horsefish

The South African horsefish (*Agripus*) sheds its skin like a snake. Most other fish gradually replace skin, as humans do.

Housefly

The average housefly lives only two to three weeks.

The principles of stabilized flight used by houseflies helped develop the first gyroscopes.

The hairs on a housefly are very sensitive to air pressure, which is why a housefly is hard to catch or kill. Fly swatters are made with holes in them to reduce the air pressure created by a swat.

Huia Bird

Now believed to be extinct, the huia of New Zealand has been described as one of the world's most remarkable birds because the male and female had different bill shapes. The bill of the male bird was strong and stout, and the female's bill was long, thin, and curved. These different bills enabled the birds to search for food together. The male tunneled into decaying logs to seek out grubs and the female inserted her bill to pull out the insect larvae.

Hummingbird

Hummingbirds are found only in the New World and take their name from the noise produced by their rapid wing beats.

The hummingbird family, Trochilidae, contains approximately 319 species.

Hummingbirds are related to the swifts and are highly adapted for life on the wing. Their feet are so weak that they are used only for perching.

In general, hummingbirds are small, fast, and very active. They range in size from the giant hummingbird (Patagona gigas), measuring eight inches, to the bee hummingbird (Mellisuga helenae), whose body averages one inch in length. Generally, the smaller a hummingbird is, the faster it must move its wings to remain airborne.

The dynamics of hummingbird flight is highly specialized to allow the birds to hover and fly backward as well as forward. The speed of forward flight is between 30 and 40 mph.

The bills of hummingbirds vary considerably, depending on the particular species adaptation for feeding. Some have short bills only one half the length of the head, while the sword-billed hummingbird has a straight bill that is as long as its head, body, and tail. In addition, some species have bills that are curved strongly downward or slightly upward. All are adaptations for feeding from specific types of flowers.

By feeding on nectar, hummingbirds are important agents of pollination.

Because of their small size and high metabolic rate, active hummingbirds feed once every 10–15 minutes.

Hummingbirds have the ability to become torpid in order to conserve energy.

The nests of hummingbirds are constructed from fibers, moss, lichen, and similar fine material and are held together by spiderwebs.

Hummingbirds are aggressive among themselves. They are highly territorial when nesting and outside the breeding season may defend feeding territories. Not only will they attack one another, hummingbirds will also attack birds that are much larger than they, like hawks and eagles. Their great speed and agility give them the advantage over larger birds.

The ruby-throated hummingbird migrates over 6,000 miles from Canada to Panama. During their migration these hummingbirds cover the Gulf of Mexico, a distance of 500 miles, in a single nonstop flight. To do so they must store up large fat reserves to provide energy for the flight.

Hyena

The spotted hyena is quite vocal, uttering a variety of calls. The most notable of these is the so-called "laugh," which is uttered during the breeding season or when the animals are excited.

Hyenas have some of the most powerful jaws in the animal kingdom, jaws capable of crushing even the largest bones of cattle and other large animals.

Some African tribes put their dead outside the walls of their villages so that scavenging hyenas will dispose of the corpses.

Although hyenas are predominantly scavengers, they can be active, aggressive hunters preying on Africa's game animals.

Hyenas are excellent swimmers and have been known to dive under water to retrieve drowned prey.

Female hyenas are larger than males. Females lead the hyena pack.

Hyrax

The hyraxes, found in Africa and the Middle East, look like large rodents or short-eared rabbits but are in fact the closest living relatives of the elephant.

The soles of a hyrax's feet are covered with thick

skin and are kept constantly moist by glandular se-
cretions. A specialized set of muscles in the foot
enables the soles to contract, forming a hollow air-
tight cup that aids their climbing swiftly over the
rocks on which they live.

The hyrax is the coney of the Bible.

Each of the hyrax's toes is capped with a tiny
hoof.

Iguana

The family Iguanidae includes about 700 species that are distributed from southern Canada to southern Argentina. Outside the New World, iguanas are found only on Madagascar and the Fiji Islands.

The Galápagos Islands are the home of the only true marine lizards in the world, the marine iguana. Like all of the large iguanas, they are herbivorous, feeding on seaweed that they reach by diving.

green iguana

When threatened, the common iguana of Central and South America uses its tail as an accurate and effective whip in defense.

Iguana tails are eaten by people in Central and South America. The meat is white and reputed to taste like frogs' legs.

Impala

Impalas are among the most graceful and swiftest antelopes. When alarmed, impalas make prodigious leaps while running, having been known to clear 30 feet in a single leap.

During the dry months impalas of both sexes congregate in large herds of several hundred animals, which later break up into smaller groups of 15–25 animals led by a male.

The largest animal ever known to have been eaten by a python was a 130-pound impala.

Insect

Insects outnumber all other types of animals combined.

Insects are consumed by humans in many areas of the world and are an excellent source of protein. Ants, beetles, caterpillars, grasshoppers, grubs, locusts, and termites are among the most popular insects eaten.

Jackal

The three species of jackals are distributed from Russian Turkestan through the Middle East and throughout Africa. All three occur together in certain parts of northeastern Africa.

Jackals travel singly, in pairs, or in small groups, feeding on any small animals they can catch. They also feed on carrion, insects, and plant material.

Jackals are members of the genus *Canis,* as are domestic dogs. The closeness of the relationship between members of this genus is evident by the successful breeding between all types of wild dogs and domestic dogs.

Jackals frequently follow lions and leopards in order to glean scraps from their kills.

Japanese Beetle

The Japanese beetle, a notorious garden pest in the northeastern part of the United States, first appeared in this country in 1916 when it was discovered in Riverton, New Jersey. It was believed to have been in larval form in the soil of potted plants imported from Japan.

Jellyfish

A jellyfish on land will eventually evaporate; nevertheless, it is able to sting for several hours after it has been beached.

The tentacles of the pink jellyfish can measure as long as 200 feet. This massive jellyfish lives in the north Atlantic.

The only species of jellyfish that has a deadly sting is the sea wasp or box jellyfish *(Chironex fleckeri)*. The sting of the sea wasp, found from Indonesia to northern Australia, causes paralysis of the heart muscle in 5–10 minutes and has caused 70 documented deaths in this century.

Kagu

Although the kagu has large wings, it flies very rarely and some ornithologists believe that it is incapable of flight. The kagu is found only on the island of New Caledonia and has been considered an endangered species since 1940, due to the depredations caused by animals introduced by humans.

Kagus feed on worms, insects, and other invertebrates but feed primarily on a type of snail that is removed from the shell by a blow from the bird's bill and then shaken to remove the shell fragments.

Kangaroo

There are three species of kangaroo inhabiting the open forests and bush of Australia and Tasmania. These are the largest marsupials, weighing up to 250 pounds and standing almost 6 feet tall.

Kangaroos, unlike most other mammals, continue to grow throughout their lives.

The scientific name of the kangaroo, *Macropus,* means "large foot." The kangaroo's hindquarters are also greatly enlarged and well muscled for leaping.

The tail of a kangaroo balances the animal when it is leaping, and it is strong enough to support the weight of the entire animal.

Kangaroos have been clocked moving at 30 mph but cannot maintain this pace for long.

Kangaroos feed at night and normally rest during the day.

Kangaroos are born after a gestation period of 30–40 days. Like all marsupials, they have well-developed nostrils, forelimbs, and large tongues, but the other external features are still embryonic. Newborn kangaroos take approximately three minutes to find the pouch after birth and are not assisted by the female. Upon reaching the pouch, the newborn attaches itself to one of the female's nipples and remains attached to it for several months. Young remain in the pouch for about 240 days, at which time they are fully developed.

Kangaroo Rat

There are 22 species of kangaroo rat inhabiting the western United States and Mexico.

Both sexes have a gland between the shoulders that produces an oily secretion and odor that may help individuals to recognize one another. Kangaroo rats must bathe in dust or the secretion from this gland will cause skin sores and mat their fur.

Kangaroo rats travel by hopping and therefore

their hind legs are greatly developed while their forelegs are reduced in size.

A kangaroo rat's kidneys are extremely efficient, an adaptation to life in the desert, which cuts down the amount of water required to survive.

Kangaroo rats feed on seeds, fruits, leaves, buds, and insects and store food against times of drought. They transport seeds to their burrows in cheek pouches.

Kangaroo rats, only a few inches tall, can jump as high as 18 inches when alarmed and are known to kick sand at a predator in an effort to escape.

Kestrel

The kestrel, known as a sparrow hawk in the United States, is one of the few birds that can hover for more than a few seconds. In England the local genus of kestrel is called the windhover.

Kestrels hunt insects on the wing, catching them with their talons. They consume only the soft bodies of their prey and discard the shell.

Kinkajou

The kinkajou, a relative of the raccoon living in Central and South America, is almost entirely arboreal and possesses a prehensile tail.

Nocturnal animals, kinkajous sleep in tree hollows

during the day and move through the trees at night, feeding primarily on fruits, though insects, birds' eggs, and small mammals are also consumed.

Kiwi

Although the kiwi of New Zealand is one of the world's most familiar birds, very little is actually known about them because of their secretive, nocturnal habits.

Kiwis are the smallest of the ratite (flightless) birds, standing about 15 inches tall, and are related to the extinct moas.

The kiwi's egg, in relation to the hen's body weight, is the largest of any bird. Kiwi eggs weigh approximately one pound, representing 12.5% of the hen's total weight. Eggs measure 5 inches by 3.5 inches in diameter.

In captivity twin chicks have hatched from a single egg.

Kiwis nest in burrows lined with feathers, grass, and twigs. Eggs are incubated by the males only and hatch after an average of 75–78 days. Kiwi chicks are fully feathered at birth and are not fed for the first 6–12 days, after which time they are able to pick up food on their own.

Kiwis feed on a variety of worms, insect larvae, and plant material, using their long beaks and strong legs to uncover food on the forest floor.

The kiwi's bill is unique in having the nostrils at

the tip. The bird's sense of smell is acute, and this adaptation, together with sensitive, specialized hair-like feathers located at the base of the bill, aids them in locating food.

The Maori called the kiwi *"Te manu huna a Tane,"* the hidden bird of Tane, god of the forest, in reference to its nocturnal habits.

Koala

Although koalas are highly specialized to feed on the leaves of about 12 species of eucalyptus trees, some authorities believe that they also eat mistletoe and the leaves of the box tree.

Despite the common name "koala bear," koalas are marsupial mammals.

The fingers and toes of the koala possess strong claws and the first and second fingers are opposable to the remaining three digits; all are adaptations for life in the trees.

Adult males usually gather a small harem of females and guard them carefully from intrusions by other males.

The koala's gestation period is 25–30 days. The young weigh about five grams at birth and remain in the female's pouch for about six months.

The koala normally does not drink water since it obtains necessary moisture from the leaves it eats. The name *koala* is derived from the aboriginal word meaning "no water."

Because eucalyptus leaves contain aromatic oils, koalas tend to have the odor of eucalyptus about them.

Only the San Diego Zoo and the Los Angeles Zoo in the United States are permitted by the government of Australia to import koalas because adequate supplies of eucalyptus can be grown for them in southern California's climate.

The koala has an exceptionally long cecum (first section of the large intestine), six to eight feet in length. This is an adaptation for gaining maximum nutritive value from the tough, leathery eucalyptus leaves.

In Australia koalas are also known as bangaroos, kollewongs, and naragoons.

Komodo Monitor

The Komodo monitor, or "dragon," is the world's largest lizard, attaining a length of 10 feet 2 inches and weighing up to 400 pounds. Although called Komodo monitors, these huge lizards are found not only on Komodo, but also on Rincha, Padar, and Flores of the Lesser Sunda Islands of Indonesia.

Komodo monitors bring down prey as large as water buffalo by severing the tendons of the hind legs with their powerful teeth and jaws.

The only Komodo monitors living in the United States are exhibited in the San Diego Zoo.

The Komodo monitor was not discovered by Western scientists until the year 1910.

Ladybug

The ladybug, a type of beetle, was imported from Australia to southern California in 1892 for the purpose of controlling mealybugs, lice, and other citrus pests.

Lark

There are approximately 75 species of lark, most of which are found in the Old World. Eighty percent of all lark species inhabit Africa.

The shape of the bill varies widely among the larks, an indication that the group feeds on a wide variety of material, from large seeds to insects. The horned lark feeds on small mollusks and crustaceans, which it finds by searching through seaweed at the tide line.

Most species of lark nest on the ground. Some species build small walls of pebbles on the exposed side of the nest as a protection against the wind.

Leech

Hirudin, an anticoagulant obtained from the salivary

serval

Western red colobus

tapir

**Abyssinian black
and white colobus**

glands of leeches, was once used in surgical operations to prevent blood clotting.

After its domestic supply of leeches was depleted in 1827, France imported 33 million for medical use.

Leeches are hermaphroditic.

Lemming

The Scandinavian lemming engages in mass migrations when its population density becomes too high. During such migrations lemmings do not hesitate to cross rivers, streams, and lakes, but many die while doing so, leading to the false notion that lemmings commit suicide.

Lemur

Five species comprise the genus *Lemur;* these primitive primates are native to Madagascar and the Comoro Islands.

Four species of lemur are highly arboreal but the fifth, the ringtail lemur, is terrestrial, living among rocks in thinly wooded country.

The breeding season of lemurs is limited. During most of the year their reproductive organs are nonfunctional.

Certain Malagasy natives will not hunt lemurs because they are thought to be reincarnations of deceased ancestors.

During the Pleistocene, a lemur, *Megaladapis,* which stood five feet tall, was found in Madagascar.

Leopard

The leopard has the greatest range of any cat, inhabiting most of Africa and Asia. Leopards inhabit all kinds of terrain from tropical rain forest to scattered forest in temperate climates to high altitudes in the area of the Himalayas.

Leopards are agile in trees and often attack prey by pouncing from above.

leopard

Leopards are primarily nocturnal.

Many leopards exhibit melanism (black coloration). Black kittens may be born in the same litter with normal-colored kittens. Black panthers are actually black leopards or jaguars.

If a leopard cannot consume all of an animal it has killed, it will cache the remains in a tree and return to feed on the carcass later.

Lion

Perhaps no other animal has played as large a role in the folklore, heraldry, and imagination of humans as has the lion. Lions formerly ranged from Africa eastward across the Middle East, southern Europe, Iran, and into India. Today lions are confined to Africa and a small population living in the Gir Forest of India.

Lions prefer open habitat and are not found in densely forested areas.

Lions are the most social of all the great cats, living in groups containing individuals of all ages and both sexes, called prides.

Lions hunt the large hoofed game animals of the African savanna. Females do the great majority of the hunting but males always feed from the carcass first, while the females and young feed after the males are sated.

Despite its reputation as king of the beasts, the lion is the second largest cat; the tiger is larger.

Lions kill their prey after stalking them to within 30–50 yards and then attacking with a short burst of speed. If a lion does not bring down its intended prey during this first expenditure of energy, it must rest, because lions tire quickly while running.

Lions are the symbol of the city-state of Venice, and the Asian city-state of Singapore is known as the "Lion City."

The earliest known pet lion was kept by the Egyptian pharaoh Ramses II around 1250 B.C. The lion's name was Antam-nekht. Antam-nekht accompanied Ramses in war, walking with the horses of his battle chariot and clawing those who came within range of his leash.

Lizard

There are approximately 3,000 species of lizards living in the world today.

Most living lizards belong to two large groups: the family Agamidae, inhabiting most of the Old World, and the family Iguanidae, found throughout the New World.

The collared lizard of the American southwest is one of the fastest running lizards, capable of attaining a speed of 16–17 mph.

Lizards of the genus *Draco,* found in Southeast Asia, have five or six pairs of specialized ribs projecting from their bodies that are covered with skin. When alarmed, these small lizards are able to extend their ribs, creating "wings," and glide as far as 50 feet.

The armadillo lizard of South America is a small slow-moving lizard with well-developed spikes on its tail. If threatened, this lizard rolls into a hoop, holding the end of its tail with its forefeet and sometimes putting the tip in its mouth. By so doing, the spikes of the tail act as a protection for its soft belly skin.

The chuckwalla of the southwestern United States escapes predators by wedging itself between rocks and inflating its body with air, making it very difficult for an attacker to dislodge it.

Llama

Llamas are well adapted for high altitude living in the Andes. The hemoglobin in their blood absorbs more oxygen than that of human blood; their red corpuscles can retain oxygen twice as long as human blood corpuscles.

As a pack animal, a mature llama can carry 80–90 pounds for 20 miles a day for 3 weeks.

Lobster

There are lobster breeding projects under way in many areas that are attempting to raise lobsters for commercial use. Some of these experiments have succeeded in raising young lobsters to one pound in 21–30 months as opposed to 7 years in nature.

Locust

Most people confuse a locust with a cicada, especially since a cicada is popularly called a "seventeen-year locust." A locust looks like a large grasshopper and moves in colonies, sometimes millions in number. Cicada look like hornets, do not travel en masse, and make those loud, distinctive noises in the trees during the summer.

John the Baptist ate locusts and wild honey during his time in the wilderness, according to the Bible. Locusts are eaten in many parts of the world, especially in the Near East, and Leviticus XI:22 permitted the Hebrews to eat locusts. They are usually roasted or fried. Explorer David Livingstone ate them in Africa and remarked that they were superior to shrimp.

A large swarm of locusts may weigh as much as 15,000 tons. Every day the swarm eats the equivalent of food for one million people and takes up 50 square miles.

The hairs on the tops of the migratory locust keep it flying. The air currents stimulate the hairs and cause its wings to beat.

Loon

All loons are primarily fish eaters, seizing their prey underwater with their bills. While fishing, they dive to depths of 30 feet and may remain submerged up to two minutes. In Europe they are known as divers.

Loons are clumsy on land and, while they are strong fliers, all but the smallest of the loons must patter over the water in a prolonged take-off run.

Both parents share in the incubation of the eggs, which lasts 26–29 days. Chicks leave the nest after one or two days, at which time they are able to swim, though they do not fly until approximately two months after hatching.

Loons have a curious territorial threat display wherein a pair will warn off other loons by means of a "plesiosaur race." This consists of both birds swimming in unison with the rear halves of their bodies submerged but with the necks extended stiffly and the bills pointed upward.

Maggot

Prior to this century maggots were widely used in medicine because when they are placed on wounds, they feed on the dead tissue and avoid the healthy tissue, allowing the wound to heal more quickly and decreasing the risk of infection.

Mallee Fowl

The mallee fowl *(Leipoa ocellata)* of Australia has one of the most complex nesting behaviors of any bird. Long before breeding commences, the male makes a hole about 1.5 feet deep in the ground and then begins to heap leaves and other vegetable material on it until a mound approximately 16 feet across and 3 feet high is formed.

Female mallee fowls may lay up to 35 eggs, at four- to eight-day intervals, into the egg pit at the bottom of the mound. By constantly attending the mound and scraping dirt on or away from the fermenting vegetation, the pair keeps the temperature of the eggs at about 92°F.

Each egg incubates for about seven weeks and chicks hatch under the soil. The chicks begin life by struggling to the surface of the mound and running into the surrounding forest, where they begin feeding. Chicks are able to fly within hours of hatching.

This unusual method of nesting requires that the parent birds attend their nest for as long as 11 months out of the year.

Mandrill

Mandrills are large, colorful baboons living in dense forests in West Africa.

Mandrills have prominent ridges on their cheeks, which in males are colored bright blue. In addition, the male's nose is a brilliant scarlet color.

Male mandrills can be formidable adversaries, weighing as much as 115 pounds and possessing massive canines, well adapted for fighting.

Manta Ray

Despite their menacing appearance and size (weighing as much as 3,000 pounds and measuring up to 20 feet wide), manta rays are harmless. They eat small sea animals like shrimp, shellfish, and plankton.

Marmoset

This family contains about 35 species of small primates inhabiting tropical forests from Panama throughout South America.

All primates possess nails on their fingers and toes but in the marmosets all of the nails except the one on the great toe have been modified and look more like claws; this is an adaptation for living in trees and moving around on large branches that are too big for them to grasp.

Marmosets tend to be omnivorous but feed largely on spiders, insects, and fruit.

Males assist in the birth of infants (normally twins are born) and it is the males who carry and tend the infants following birth. Infants are transferred to the female only at feeding times.

The word *marmoset* dates to Middle English and was adapted from the Old French word *marmouset*, meaning "manikin" or "grotesque image."

Mayfly

The mayfly usually lives only 24 hours as an adult. After spending one to three years as a naiad, it becomes an adult, during which time it molts twice, mates, and lays eggs.

Meadowlark

Although the meadowlark sings like a lark, it is not a true lark, but a member of the blackbird family.

The western meadowlark, usually the bird heard on the soundtracks of Hollywood westerns, was discovered 80 years after the eastern meadowlark. It was given the species name *neglecta* in 1844 because of this.

Mexican Hogfish

The Mexican hogfish starts off life as a female and then becomes a functional male, a transformation characteristic of many members of the wrasse family.

Mink

Mink are close relatives of skunks and also produce a foul-smelling fluid when they are annoyed. The mink's original scientific name was *putorius,* meaning "foul smelling."

Mink are agile and aggressive hunters and will attack animals many times their own size.

Moa

The now extinct moas of New Zealand were the tallest birds that ever lived. Though moas did not weigh as much as the elephant birds of Madagascar, the largest moa *(Dinornis maximus)* stood as tall as 12 feet.

The order *Dinornithiformes* (giant birds) contained an estimated 19 species. Most were ultimately exterminated by the human inhabitants of

New Zealand who hunted them for meat and eggs.

Although it is not known precisely when the last moa became extinct, it is believed that some species survived until the nineteenth century.

Kiwis are regarded as the sole survivors of the moa group.

Mockingbird

Mockingbirds *(Mimus polyglottos)* range throughout the central, eastern, and southern United States and into Mexico.

Mockingbirds have the ability to imitate other species of birds, but some scientists believe that only about 10% of their calls are cases of true mimicry.

Mole

The mole is a voracious eater of worms, grubs, and insects. The average mole eats one third to two thirds of its own weight daily.

Mongoose

Famous for its ability to kill cobras, the mongoose is kept in houses in some parts of the world to help eliminate rodents and snakes.

The word *mongoose* is from the Prakrit word *manguso,* which means "ferretlike animal."

Mongooses were imported to the island of

Jamaica to eliminate rats in the cane fields and have become a case of biological control gone wrong. The mongooses are destroying much of the island's fauna, and the rats, being survivors, are still around.

Monkey

Monkeys are divided into two groups: the *Ceboida,* representing the monkeys of the New World, and the *Cercopithecoidae,* the monkeys of the Old World.

There are about 100 species of monkey living today.

The Barbary macaque inhabiting the Gibraltar Peninsula is the only species of monkey found in Europe.

Only a few species of monkey from the New World have prehensile tails. This adaptation is not found on Old World monkeys.

New World monkeys normally have 36 teeth, while Old World monkeys have 32.

Although most species of monkey live in tropical or semitropical environments, the Japanese macaques inhabit the cold northern regions of Japan, often living along the coast, and are often called snow monkeys.

Monkeys, especially those of the genus *Macaca,* have played an important role in biomedical research. For example, the Rh blood factor was first demonstrated in rhesus macaques, and the Philip-

monkey

pine, or crab-eating, macaque was used extensively in the development of the polio vaccine.

One of the strangest-looking of all monkeys is the proboscis monkey, a leaf-eating species that only inhabits the swampy mangrove forests of the island of Borneo. The nose of the males continues to grow, and in old males the nose may measure as much as three inches in length. Proboscis monkeys are excellent swimmers and their toes are partially webbed to aid them in swimming.

Moose

Moose are the largest members of the deer family. Bulls can stand 6½ feet at the shoulder and weigh as much as 1,800 pounds.

Moose are found in the wooded areas of Alaska, Canada, the northwest United States, Norway, Sweden, and eastward across the Soviet Union to Manchuria and Mongolia. Moose prefer wet areas with abundant stands of willows and poplars, on which they feed.

Moose also feed on water vegetation and may submerge themselves entirely to reach vegetation on the bottom of lakes.

In Europe the moose is known as the elk.

Mosquito

Only female mosquitoes bite people; the males are adapted for feeding on pollen, nectar, and plant juices.

Some species of mosquito inhabit the Arctic. The pitcher-plant mosquito actually freezes in the ice on the leaves of plants and resumes activity after the spring thaw.

As carriers of the malaria-causing germ plasmodium, mosquitoes have caused more human deaths than all wars in history.

Moth

Many species of moths do not eat as adults and their sole function is to reproduce.

The hawkmoth of the tropics has a ten-inch proboscis that is used to suck nectar from flowers.

The male gypsy moth can detect a female of its own species up to seven miles away.

The great owlet moth of Central and South America has the largest wingspan of any moth: 14.17 inches.

The Mexican jumping bean is the larva of the *Carpocapsa* moth living inside the seed of the Mexican shrub of the genus *Sebastiana*.

Mouse

The North American meadow mouse produces the greatest number of litters for a mammal per year: up to 17.

Musk-ox

The distribution of the musk-ox is limited to Alaska, northern Canada, and Greenland.

For protection against cold, the coat of the musk-ox has two parts: coarse, dark-brown guard hairs that reach nearly to the ground, and an inner coat of fine, soft light-brown hair called qiviut that is so dense that it cannot be penetrated by cold or moisture.

Musk-oxen live in herds of up to 100 individuals. When attacked, musk-oxen form a defensive circle with the calves inside. This behavior is an adequate defense against attacking wolves but proved disastrous with the advent of firearms, for it is an easy matter for "sportsmen" to kill an entire herd, since the animals remain in their defensive posture and do not attempt to escape.

Myna

Famous for its ability to mimic voices, the myna bird is a member of the starling (Sturnidae) family inhabiting India.

Narwhal

Narwhals are small whales inhabiting Arctic seas. Narwhals possess two teeth in the upper jaw; in males one of these teeth (usually the left one) develops into a huge, spiraling tusk that may be as long as eight feet or one half the length of the narwhal's body.

The function of the narwhal's tusk is unknown. It has been suggested that the tusks are used for breaking ice, for fighting, or for feeding, but none of these explanations seems correct. Many scientists feel that the narwhal tusk is a specialization that has evolved beyond the point of usefulness.

In medieval times narwhal tusks were highly prized because they were thought to be unicorn horns. They first appeared in Europe as Norse trade items from Greenland and Iceland.

Nautilus

The chambered nautilus, a kind of mollusk found in the South Pacific and Indian oceans and bearing a spiralled, chambered shell, is one of the oldest types of animals still living, having been in existence over 450 million years.

Northern Manatee

The northern manatee, or Steller's sea cow, was a relative of the manatees and dugongs that lived around the Bering Islands and adjacent regions of the northern Pacific.

The Steller's sea cow was huge, measuring 27–35 feet in length and weighing as much as 4 tons. Never numerous, Steller's sea cows were completely exterminated within 50 years of their discovery by sailors hunting them for food.

Numbat

The numbat, a small marsupial living in southwestern Australia, is unusual because the females lack pouches. In the region where the pouch would normally be found, females have four teats, and when young are attached to the nipples, only the long hairs of the female's underbelly protect them.

Captive numbats have been known to eat 10–12 thousand small termites in a day, swallowing them whole.

Octopus

Like a chameleon, the octopus can change colors. When it is excited or angry, it turns red, and when frightened, it turns white. It also can turn blue and green to blend in with its background.

To escape from predators an octopus can let an arm break off and a new arm will eventually grow in its place.·

The octopus is an advanced shellfish that has lost the need for a shell because of its mobility.

Like squid, octopuses are able to eject ink as a defensive measure to confuse predators.

Octopus eyes have been used in research experiments because they closely resemble the human eye. Octopuses have also been taught to read letter-like shapes.

Oilbird

The oilbird of northern South America, a cave-dwelling bird, uses a sonar system similar to that of a bat. It emits clicks that bounce back and alert the bird to the presence of obstacles. The cave swiftlet of southern Asia is similarly equipped.

Okapi

Okapis are the closest living relatives of the giraffe, being classified as members of the family Giraffidae. They inhabit the eastern equatorial rain forests of the Congo.

Truly beautiful animals, okapis are covered with short, sleek chocolate-brown hair. The rump and forelegs are transversed with white strips of various widths, and the facial markings are light. They have large dark eyes and large ears, and the males have small skin-covered horns.

The tongue of the okapi is so long that the animal uses it to wash and clean its eyes.

In the wild, okapis are extremely wary and will retreat through the forest at the first hint of danger. They especially fear humans, leopards, and snakes.

The existence of the okapi was unknown to Western science until 1900 when Sir Harry Johnson described the animal on the basis of two skins and one skull.

The first white man to hear of the okapi was Henry Morton Stanley during his exploration in search of Dr. David Livingstone. Though Stanley never saw an okapi, he was told about them by the Mbuti pygmies of the Ituri forest.

Opossum

The opossum is remarkable in that it survives in the United States at all. Marsupials generally cannot successfully compete with placental mammals, but opossums are so adaptable that they compete with placentals extremely well.

The American opossum has one of the shortest gestation periods of any mammal: 12–14 days. At birth an opossum weighs about one fifteenth of an ounce and immediately makes its way to the female's ventral pouch for a ten-week nurturing period. The adult opossum weighs 2,000 times what it weighed at birth.

The opossum is the only North American mammal with a prehensile tail. It uses the tail occasionally to help support itself.

Opossums don't always "play possum" when threatened. In attempts to defend themselves they can be quite lively. Sometimes they do enter a catatonic state when surprised, but this is an involuntary reaction. An opossum has an extremely small brain and in the face of danger its brain jams and the animal goes into a coma-like state.

The part of the brain that regulates body temperature is not functional in young opossums; infants are kept warm by their mother's body heat.

Orangutan

Orangutans, the great apes of Asia, are found only

orangutan

on the islands of Sumatra and Borneo. Today fewer than 5,000 orangutans survive in the wild.

Among primates, only the gorilla is larger than the orangutan; adult males may weigh as much as 260 pounds.

Compared to gorillas and chimpanzees, orangs do not have large brow ridges, but adult males have well-developed cheek pads and the Bornean orang has an additional well-developed throat pouch.

The least social and most arboreal of the great

apes, orangutans move through the trees of the forest singly or in pairs. They are active during the day and normally move about unhurriedly.

During rainstorms orangs have been known to cover themselves with large leaves. At night they build sleeping platforms of sticks and vines.

The word *orangutan* is Malay, meaning "people of the forest."

An orangutan at the Philadelphia Zoo named Guas lived to be 57 years of age. It died in February 1977.

Oriole

Orchard orioles build two nests, one for the female and the eggs and another for the male.

Ostrich

The ostrich is the largest living bird, standing up to 8 feet tall and weighing as much as 340 pounds.

Ostriches have only two toes on each foot, an adaptation for running and walking that gives greater strength and thrust to the foot. Ostriches can reach speeds of up to 40 mph.

Until the first decades of the twentieth century, ostriches were common throughout Africa and the Middle East. Today they are found only in eastern and southern Africa. In southern Australia some domesticated ostriches have become feral.

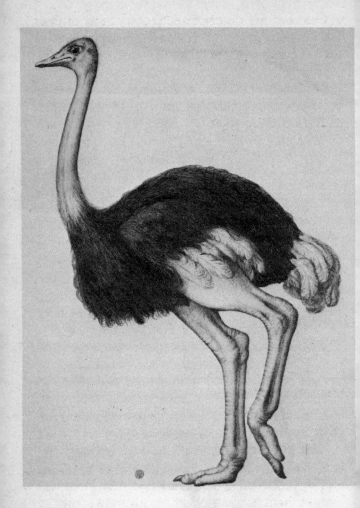

ostrich

Ostriches are omnivorous, though most of their food is plant material. They swallow quantities of grit and stones to aid digestion by grinding their food.

Ostriches are polygamous and nest communally. Females deposit their eggs in a shallow pit; clutch sizes vary from 15 to 60 eggs, with each female laying 6 or 7 eggs. Although ostrich eggs are large, they represent only about 1.5% of the weight of the female, an unusually low figure for a large bird.

Both sexes incubate the eggs, males incubating at night and females during the day. A large percentage of eggs fail to hatch.

Ostriches have been farmed for their plumes in South Africa since the 1850s. In the early part of the twentieth century there were over 700,000 captive ostriches on farms in Africa, the United States, Europe, and Australia. Today there are only about 25,000 ostriches on farms, principally for the production of high-grade leather.

Ovenbird

The rufous ovenbird builds a domed two-room nest consisting of an antechamber and a grass-lined nesting chamber where the female deposits her eggs. The nest is oven-shaped, a foot in diameter, and weighs about nine pounds.

Owl

All owls have soft, dense plumage that allows them to fly without making any sound, an extremely useful adaptation for a nocturnal predator.

Female owls tend to be larger than males, though the coloration of the plumage is similar in both sexes.

Owls kill their prey by grasping them with their strong talons and crushing the base of the skull with the beak.

snowy owl

An owl's eyes have overlapping fields of vision. The retinas of owls are extended and have a heavy concentration of sensitive cells. These adaptations give owls remarkable visual acuity and excellent night vision.

Owls also have superb hearing; the ear apertures often occupy the whole side of the skull. In many species the ear apertures are asymmetrical, helping the owl to hone in on sounds from great distances. Owls are especially sensitive to high-frequency sounds such as those made by small rodents.

Owls range in size from the eagle owl of Scandinavia, which weighs approximately 9 pounds (28 inches), to the elf owl (5.5 inches, .9 ounce) of the southwestern United States, which feeds mainly on insects, spiders, and scorpions.

Owls consume all parts of the animals they prey upon and between feedings regurgitate pellets containing feathers, bones, teeth, and other indigestible material.

Ox

A furlong was originally defined as the distance that oxen could pull a plow without having to rest.

Palm Chat

The small palm chat bird of the Caribbean builds communal nests that may measure ten feet in height and be four feet in diameter. Up to 30 pairs of palm chats may live in a given nest.

Panda

Pandas are native to four provinces of China, where they inhabit high, rugged bamboo and rhododendron forests.

Because of its striking black and white coloration and bearlike body shape, the panda is immediately recognizable and has been adopted as the symbol of the World Wildlife Fund.

Pandas may be aberrant bears, as many zoologists believe. Their heads are massive, partially as a result of expanded zygomatic arches (cheekbones), large jaw muscles, and massive teeth, all of which are necessary for chewing the leaves and stems of bamboo. Many other zoologists, however, believe that pandas are more closely related to raccoons.

Pandas have an unusual modification of the carpal bones in the wrists of their forepaws. These bones have evolved into a functional "thumb" and allow pandas to grasp the bamboo on which they feed.

In the wild, giant pandas are solitary except during the breeding season. They do not hibernate, but remain active throughout the year.

Pandas spend 10–12 hours a day feeding on grasses, irises, crocuses, and occasional small rodents, as well as bamboo.

Panda cubs, weighing about 4 pounds, are usually born in January and gain an average of 2½–3 pounds per month during their first year. Panda cubs are entirely white at birth.

The first panda seen alive outside of China was exhibited at the Chicago Zoological Park in 1936.

Panther

A panther is not a specific animal but a catchall word for many members of the cat family. The genus *Panthera* includes the lion, tiger, leopard, and jaguar.

Parrot

All parrots are members of the family Psittacidae. The members of this family display a variety of forms, from the large macaws of South America to the tiny hanging parrots of Asia, which feed and sleep while hanging suspended from branches.

In general, parrots are brightly colored, have

macaw

stout hooked bills, short necks, plump bodies, and rounded wings.

Parrots are found primarily in tropical regions, though some species have adapted to temperate environments like that of New Zealand.

Until the twentieth century one species of parrot was found in the United States, the beautiful Carolina parakeet *(Conuropsis carolinensis),* which measured about a foot long and had a yellow and orange head and green body plumage. The last known living specimen died in the Cincinnati Zoo in 1914 just one month before Martha, the last known passenger pigeon, died in the same zoo.

Most parrots feed on seeds, fruits, and vegetable matter, but some, like the lories and hanging parrots, have adapted to feeding on nectar, pollen, and fruit juices. Most parrots hold their food in one foot while feeding.

The vast majority of parrots nest in holes in trees. All parrots lay white eggs without markings that are slightly spherical in shape.

Some parrots are able to mimic human speech and other sounds because of the anatomical construction of their bills. Among the best "talking" parrots are the African gray parrot and the yellow-naped Amazon parrot.

Peacock

Until the wild turkey was brought from Mexico to Europe, the peacock was a table bird in Europe.

Pelican

Pelicans are an ancient bird. The first identifiable pelican was found in France and dated to the Miocene era, 20 million years ago. Pelicans are highly adapted for swimming and flying but are ungainly on land.

Depending on which authority is followed, there are six to eight species of pelican. All are white with the exception of the brown pelican of North America. The brown pelican is the only species that is almost exclusively marine in its habits.

The flesh pouch on the pelican's bill is not used to store fish but rather as a scoop to trap them. The brown pelican is unique in that it makes spectacular dives into the sea to catch its prey. Other species swim across the water, darting their heads at fish as they pass by.

Pelicans are among the largest flying birds, some species attaining a weight of 25 pounds and having wingspans of up to 9 feet. Their skeletons, however, are extremely light since the bones are infused with many air sacs. The skeleton of a large pelican weighs less than 2 pounds.

Pelican chicks are not able to fly for 60–70 days after hatching. Before they are able to fly, pelican chicks become much heavier than their parents. Adults usually stop feeding their young before they can fly and the extra weight ensures their survival until they can find food on their own.

Penguin

The name penguin was originally applied to the extinct great auk that inhabited the North Atlantic. The scientific name of the great auk was *Pinguinus impennis.*

Penguins are highly specialized seabirds found only in the Southern Hemisphere. The largest is the emperor penguin, which stands 3 feet tall and weighs 65 pounds, and the smallest is the little blue penguin of Australia and New Zealand, which stands one foot tall and weighs 2–2½ pounds.

Unlike other flightless birds, the penguin has a strongly keeled breastbone and powerful breast muscles, as do birds that fly. This is because penguins literally "fly" through the water, propelling themselves with their wings.

The feathers covering a penguin's body are short, curved, and have dense down at the base. The tips of the feathers overlap like roof tiles, and many glands in the skin produce oil that creates a tight, warm, waterproof coat, preventing excessive heat loss.

Penguins are highly social birds usually found in large groups both on land and at sea. They breed communally, the nesting colonies sometimes containing several million pairs.

When swimming, penguins "porpoise," allowing them to breathe without reducing the swimming speed, normally 5–10 mph.

Penguins have fleshy backward-pointing spines lining their mouths and on their tongues that help them to handle slippery prey. In addition, all penguins have strong bills with sharp cutting edges.

Penguins drink both fresh and salt water; a special salt gland above the nostrils removes the excess salt from their bodies. Polar species eat snow for water.

Among emperor penguins, only the males incubate eggs, holding them on their feet for two months and covering them with a brood pouch that keeps the eggs warm in temperatures down to −57°F.

The emperor penguin has square pupils.

Great care must be taken to prevent respiratory

diseases when polar penguins are brought into captivity. Because their normal Antarctic environment is virtually germ-free, polar species have little or no immunity to common germs.

The male Adélie penguin may lose up to 40% of its body weight while incubating its egg.

Phalarope

The red-necked phalarope female is larger, more colorful, and more dominant than the male, who incubates the eggs and cares for the young. The female selects the nesting site and defends it against other females.

Pheasant

Pheasants have a large distribution, being found over most of the world. They don't live in the Arctic, Antarctic, certain parts of South America, and most oceanic islands.

Zoologically, quails are grouped with the pheasants.

The domestic chicken, now probably the most numerous bird in the world, is descended from the red jungle fowl (all chickens are members of the pheasant family) of southern Asia.

Only one species of pheasant is found in Africa, the Congo peacock, which was not discovered until 1936. Most species of pheasant are natives of Asia.

white-eared pheasant

Most pheasants are polygamous and display re-markable sexual dimorphism (visible sexual differ-ences). Males defend large territories inhabited by several females. The males breed with each female and play no role in nesting or tending the young.

The pheasant's polygamous breeding behavior increases competition between males, leading to the evolution of such bizarre and ornamental structures as the peacock's spectacular tail.

Members of the pheasant family probably have more economic importance to humans than any other family of birds because of the role played by the domestic chicken in providing eggs and more efficiently converting food to meat than any domestic mammal.

Pig

Pigs are the oldest and most generalized members of the order Artiodactyla, the even-toed hoofed animals. They have remained unchanged because they have a generalized body form and are superbly adaptable omnivores.

There are 5 genera of wild pig, with approximately 8 species and 76 subspecies.

In most species of pig, males grow large tusks, which are kept razor-sharp by constantly rubbing the upper and lower tusks against one another. The most spectacular tusks of any pig are those of the babirusa from the Celebes. The upper tusks grow through the skin of the snout and grow in a curve toward the eyes, while the lower tusks grow straight up out of the mouth and follow the same arc as the upper tusks. Of no possible practical use, the babirusa's tusks are sexual adornment.

All pigs are excellent swimmers, crossing rivers and swimming to islands that are many miles away from the mainland.

The bearded pigs of Borneo sometimes migrate in large herds. In 1954 Dyak hunters ambushed and killed so many pigs that a river was fouled with pigs' blood and Muslims downstream threatened war.

Pigs eat almost anything, including roots and tubers, which they locate with their extremely acute sense of smell. This characteristic is employed by the French, who use pigs to locate the delicious wild truffle.

Male European wild boars have heavy shields of thick, heavily haired skin on both sides of their bodies for protection from tusk wounds during mating season fights with other boars. There is an old Indian adage that states that wild pigs are so ferocious they are the only animals that can "drink at a river between two tigers."

Domestic pigs are efficient food converters; for every 100 pounds of food intake, a pig produces 20 pounds of meat, as opposed to 7 pounds of meat for a domestic cow.

Pigs are highly intelligent and easier to train than dogs.

A pig's heart and teeth are very similar to those of humans. As a result, pigs are widely used in heart research and the valves from the hearts of pigs are commonly used in human heart valve transplants.

The largest wild pigs are the giant forest hogs of Africa, which can be 6 feet long and weigh six hundred pounds; the smallest are the pygmy hogs of India, which are ten inches across the shoulder and two feet long.

Pigeon

There are approximately 225 species of pigeons and doves inhabiting most parts of the world.

As a group, pigeons are quite diverse in size and coloration, some of the Old World fruit doves being among the most colorful birds.

All types of domestic pigeon are derived from the European rock dove *(Columbia livia)*, which nests on rocky ledges in the wild, one reason why feral pigeons commonly flock around buildings.

Most pigeon species are gregarious; the extinct passenger pigeon *(Ectopistes migratorius)* associated in flocks sometimes numbering over a billion birds. The great naturalist John Audubon once recorded a flock that took three full days to pass over one spot in Kentucky.

Most pigeons build flimsy nests of twigs in trees. When the young first hatch, they are fed on a substance known as pigeon's milk, a curdlike substance secreted by specialized cells lining the parent birds' crops.

Pigeons drink in an unusual manner for birds: Instead of elevating their bills full of water to let it run down their throats, pigeons immerse their bills and suck up water, a habit also shared by sand grouse and button quail.

The dodo of Mauritius, a worldwide symbol of extinction, was a flightless, ground-dwelling pigeon.

Piranha Fish

Not all piranha fish are vicious flesh-eaters. The *pacu* species, for example, are plant-eaters. Piranhas also have a reputation for being a small fish, but some species, like *Serrasalmus piraya,* grow as large as two feet in length.

The name "piranha" or "tooth fish" was given to these fish by South American Indians who used the piranha jaw as an all-purpose tool.

Platypus

When the duckbill platypus of Australia and Tasmania was first brought to the attention of European zoologists in the eighteenth century, the stuffed specimens were considered to be frauds. The platypus is, in fact, the ultimate in mammalian incongruity, possessing a broad, leathery snout resembling a duck's bill; a flattened, beaverlike tail; webbed feet, and spurs on the hind legs of males, which they use in defense to inject a toxic poison. As if all this weren't enough for the scientists to cope with, female platypuses lay eggs rather than give birth to live young.

As adults, platypuses have no teeth, but rather, hornlike plates running the length of both jaws. At the front these plates have sharp ridges but toward the back they are flattened for crushing; they grow continuously.

Females lay one to three sparrow-size eggs in a nest of leaves constructed in an elaborate burrow, which the female builds on a riverbank. During the seven-to-ten-day incubation period the female lies with her body curled around the eggs to provide warmth. The young hatch blind and naked and remain in the nest for a period of about four months.

Platypuses feed mainly on stream bottoms, using their bills to probe for crayfish, shrimp, insect larvae, tadpoles, snails, worms, and small fish.

Females have no nipples but secrete milk through pores in their abdomens.

Poorwill

Hibernation is rare in birds but the poorwill, a small desert bird found in the United States, does hibernate. Just before hibernating, the bird acquires a thick layer of fat.

Porcupine

Porcupines inhabit both the Old and New Worlds. There are approximately 43 species. In the Old World, porcupines of the family Hystricidae range from southern Italy to Africa and eastward to China, Southeast Asia, and the Philippines. In the New World, porcupines range from the Arctic Ocean southward to South America.

Porcupines are among the largest rodents, feeding on a wide variety of plant material and occasionally on carrion.

Porcupine spines are stiffened, sharpened, and thickened hairs that in some of the Old World species may measure up to 15 inches in length. The quills of some species have minute barbs. Contrary to popular legend, porcupines cannot shoot or throw their quills.

In general, Old World porcupines are terrestrial animals, while New World porcupines are adapted for an arboreal existence.

Porcupines are good swimmers, their hollow quills furnishing considerable buoyancy.

Young porcupines are well developed and rather large at birth. They are born with long quills that are soft and gradually stiffen in the weeks following birth.

Porpoise

The porpoise is in the whale order (Cetacea), as are the dolphins. They are both in a suborder of toothed whales.

Portuguese Man-of-War

This stinging animal is not an individual organism but a colony of interdependent animals, each with specialized functions. The float (pheumatophore) keeps the colony together and afloat; the feeding polyps (gastrozooids) gather food; the tentacle polyps (dactylozooids), up to 20 feet long, contain the stinging cells; and the reproducing cells (gonozooids) enable the colony to reproduce.

Praying Mantis

Early Muslims thought that praying mantises "prayed" while facing toward Mecca.

Large species of praying mantises will eat small frogs, lizards, and birds.

Pronghorn

The pronghorn of western North America is the only member of the family Antilocapridae.

Pronghorns are not true antelope but the only surviving members of a group of ungulates that evolved in North America. Both sexes carry horns that consist of permanent forked, bony cores covered by a leathery sheath of compressed hairs that is shed annually after the breeding season. The new sheath grows upward under the old sheath before the old sheath is shed.

Without doubt, pronghorns are the swiftest animals inhabiting North America, being able to attain speeds of over 40 mph, though their normal cruising speed is around 30 mph. After a fast run the pronghorn shakes its body; the reason is unknown.

Bucks fight to build up harems in late summer; the average harem consists of 15 females.

Young pronghorns develop very quickly on milk that is extremely rich in solids. Within three months of birth young pronghorns are almost as swift as adults and have attained adult coloration.

Pteranodon

The pteranodon, the largest flying creature that ever lived, had a wingspan of around 27 feet.

Puffin

The puffin, a type of auk, actually flies underwater much like a penguin does. The puffin, however, is also able to fly through the air. Some species of puffins can go 240 feet in 2 minutes underwater.

Puma

The puma *(Felis concolor)*, also known as the cougar or mountain lion, is the largest member of the genus *Felis*, which also includes the domestic cat.

The puma has the largest range of any native mammal in the New World, occurring continuously from British Columbia in Canada to Patagonia in South America.

As a result of its enormous range there are many races and local varieties of puma and therefore significant size differences within the species. Adult

puma

pumas range in body and tail length from just over 5 feet to just over 9 feet and weigh 75–230 pounds.

The puma is one of the two cats in the Americas that does not have a spotted coat as an adult, though kittens are spotted.

Pumas feed on a variety of animals, preying on anything they can subdue and kill. In general, they will not return to a kill after feeding.

Pumas have tremendous agility in trees and can jump great distances.

Python

The reticulated python of Southeast Asia is the world's largest snake, attaining a maximum length of around 30 feet and weighing over 300 pounds.

A python kills its prey by constriction, encircling the animal's body with several coils of its heavy body and squeezing until the prey is suffocated.

Pythons generally swallow their prey head first.

When threatened, the ball python coils itself into a tight ball to protect its head.

Quagga

The quagga was a small zebra, closely related to the Burchell's zebra, which inhabited southern Africa until the middle of the nineteenth century. The color pattern of the quagga was unusual in that striping was only present on the head, neck, and shoulders. The rest of the body was dark and the legs and belly were lighter.

Due to relentless hunting for their hides and meat, quaggas were exterminated in the wild by 1860. The last known quagga, a mare, died in the Artis Zoo, Amsterdam, in 1883.

Rabbit

Rabbits, hares, and pikas belong to the order Lagomorpha and occupy a wide variety of habitats on all of the continents except Antarctica and Australia (they have been introduced in Australia).

Though resembling rodents in some ways, rabbits are not related to the rodents and show more similarities to certain types of hoofed mammals.

Members of this order have a remarkable adaptation for deriving the maximum nutritive value from food. Their fecal material is of two types: moist pellets that are eliminated and later eaten, and dry pellets that are not eaten. Thus food passes through the digestive tract twice.

The main difference between rabbits and hares is that rabbits are born naked, blind, and helpless, while hares are born fully developed and able to care for themselves.

The jackrabbit can attain speeds of over 45 mph for short distances.

The country of Spain was originally known as *tsapan,* the Phoenician word for rabbit, because so many of the animals were found there.

The Cape or European hare is the original Easter bunny. In German legend Eostre, the goddess of spring, created the hare from a bird and in gratitude the hare laid eggs in the bird's honor during the Easter festival.

The volcano rabbit of Mexico is an extremely rare animal only found at elevations of 300–3,600 meters on the slopes of volcanoes outside of Mexico City. Instead of hopping, volcano rabbits trot.

Raccoon

Raccoons exhibit a high intelligence and manual dexterity that rival those of the monkeys. Raccoons can turn on faucets, unscrew jar lids, open trash cans, and teach their young to do the same.

In the southern part of their range (southern Canada, U.S., Mexico, and Central America), raccoons are active throughout the year, but in the north they sleep during the winter after fattening up during the fall months.

Raccoons are omnivorous but prefer aquatic animals such as frogs, fish, and crustaceans. They also consume a variety of plant material.

Captive raccoons regularly wash their food. In the wild, raccoons do not always exhibit this behavior, but when they do so it is to remove sand and grit from food.

Rat

An adult rat can:

squeeze through a hole only one inch in diameter (its skull and skeleton are extremely flexible);

fall five stories and not get hurt;

gnaw through cinder block, lead pipes, underground cables, and sheet aluminum;

survive radioactive fallout of an atomic bomb (as they did on the Bikini atolls).

Rattlesnake

Rattlesnakes hunt at night, seeking their prey with the aid of heat-sensitive pits located in front of their eyes. These specialized organs are sensitive enough to detect temperature changes as small as $1/1000°$ F.

Rattlesnakes are viviparous animals (giving birth to live young).

It is often believed that the number of rattles on a rattlesnake will indicate its age in years, but this is not the case. A new segment of the rattle is formed each time the rattlesnake sheds its skin, and rattles are frequently broken.

The venom of the rattlesnake differs from that of the cobra. Rattlesnake venom destroys tissue (hemotoxic), while that of the cobra attacks the nervous system (neurotoxic).

Raven

Ravens are the largest of the passerine (perching) birds.

Ravens and their kin, the crows and jays, display a highly developed mentality; many scientists believe that the family (Corvidae) to which these birds belong contains the most highly evolved birds. This contention is supported by the high degree of adaptability shown by these birds and by their performances in laboratory tests.

Ravens are found in both North America and Europe and range into the Arctic, where they are a common sight around garbage dumps.

Ravens are primarily omnivorous scavengers, though they will kill small and sick animals for food. They will eat almost anything that they can swallow.

Ravens are monogamous and pair for life. They build large nests in trees and lay 2–4 eggs, which the female alone incubates for 16–21 days.

Because ravens begin incubation when the first egg is laid, the young hatch over a period of several days. During times of scarce food supply the younger, smaller chicks may die because they are not strong enough to compete for food with their older siblings.

Ravens were taken along by Vikings on their westward explorations. Periodically a raven would be released and if it returned to the east the Vikings would continue westward. If, however, the raven flew in another direction, the Vikings would follow it on the assumption that the raven was heading for land.

Reptile

There are approximately 5,175 living reptile species divided into four orders: the Rhynchocephalia, the tuatara; the Squamata, lizards and snakes; the Crocodylia, crocodiles, alligators, and gavials; and the Chelonia, turtles, tortoises, and terrapins. The order Squamata contains the largest number of living species with over 90% of the reptiles.

The reptiles are defined as cold-blooded air-breathing, vertebrate animals that are covered with scales or bony plates.

Rhea

The rhea of South America is the largest bird inhabiting the Americas. The larger of the two species, the common rhea, stands 5 feet tall and weighs up to 55 pounds.

The common rhea inhabits the open country of the pampas, while the smaller Darwin's rhea is found in the Andes at elevations of up to 16,000 feet. Both types live in small bands of 5–20 individuals.

Like ostriches, rheas feed primarily on vegetable materials but will eat insects, lizards, and small mammals if the opportunity arises.

Adult males fight for females and territory at the start of the breeding season, using their beaks, necks, and powerful feet as weapons.

After a male gathers a harem of females, he constructs a nest by scooping out a small depression in the earth, then leads the females to the nest, where they literally line up to lay. Each female may lay as many as 15 eggs.

The male alone is responsible for incubation and vigorously defends the nest even against his own females, who wish to lay more eggs. After a six-week incubation period, the male looks after the chicks for four or five months, at which time they are almost full grown.

Rheas are swift runners, possessing only 3 toes on each foot. They may attain speeds approaching 40 mph.

Rhinoceros

The five living species of rhinoceros inhabit savannas, scrub forests, and dense jungle in Asia and Africa. All are fast approaching extinction because they are slaughtered for their nasal horns that are erroneously believed in Asia to have aphrodisiacal properties.

The living rhino bear one or two horns on their heads, which are composed of keratin (the same material as fingernails) and which are not permanently attached to the skull. The longest recorded

black rhinoceros

rhinoceros horns belonged to an African white rhinoceros and measured 62¼ inches for the front horn and 53½ inches for the back horn.

During the 1970s poachers killed 90% of the rhino living in Africa. Today most of the rhino in game parks are provided with armed guards 24 hours a day.

In general, rhino are solitary except for the white or square-lipped rhinoceroses, who associate in groups of both sexes and all ages.

Rhino have poor eyesight, but their senses of hearing and smell are acute. Leopards, lions, and hyenas prey on young rhino, but adults have no enemies other than humans.

Rhino deposit their dung in piles that serve as territorial markers.

The skin of the Indian rhinoceros *(Rhinoceros unicornis)* and the Javan rhinoceros *(R. sondaicus)* differs from the African species in that it has a number of loose folds, giving the animals the appearance that they are wearing plate armor.

Indian rhino defend themselves with long sharp lower tusks rather than with their horns.

The smallest species of rhino is the Sumatran rhino *(Dicerorhinus sumatrensis)*, which measures about nine feet in length and weighs just over one ton. Even as adults these rhino are covered with long reddish-brown hair.

The largest rhinoceros species is the Northern white or square-lipped rhino *(Ceratotherium simum cottoni)*, which may measure 16 feet in length and weigh over 4 tons.

Roadrunner

Roadrunners *(Geococcyx californianus)* inhabit dry regions from the southern United States southward into Central America. They are one of the best-known members of the cuckoo family, Cuculidae.

The diet of the roadrunner consists of a variety of plant and animal material, but they prefer lizards and snakes. Roadrunners are adept at killing rattlesnakes.

Robin

The American robin *(Turdus migratorius)* is actually a variety of thrush.

potto

peacock

jaguar

jaguar cub

Female robins are completely responsible for nesting and the incubation of the eggs.

Rodent

The rodents are the largest order of mammals. Within the order there are 35 families, approximately 351 genera, and over 6,400 living species.

The distribution of Rodentia is almost worldwide and their adaptations are so varied that they inhabit most types of habitat. Despite their diversity, rodents have some remarkably similar structural characteristics that identify them as members of this order. For example, the incisor teeth of all rodents are chisellike and grow continuously throughout the animals' lifetimes.

Some rodents have cheek pouches, which open near the angle of the mouth. External pouches are lined with fur and can be turned wrong side out and cleaned.

Some rodents have tails, which break off easily like those of lizards so that if they are caught by the tail they can escape. The broken tail will be partially replaced by regeneration.

Sand Grouse

Sand grouse, birds inhabiting the arid regions of Africa and Eurasia, belong to the same order as the pigeons and doves.

Living primarily on the ground, sand grouse are cryptically colored to blend in with the rocks and sand on which they live.

Sand grouse must drink every day and may fly long distances to water holes, where they congregate in flocks sometimes numbering in the thousands.

Sand grouse chicks are precocial, which means they are able to feed themselves immediately after hatching. They cannot fly for five or six weeks after hatching, however, and find it hard to obtain water. For this reason, the male sand grouse has specialized belly feathers that soak up and store water. Males can transport water to their chicks from as far as 20 miles away. Chicks drink by nibbling at the water droplets held by these feathers.

Sapsucker

Unlike other woodpeckers, the sapsucker's tongue does not extend past its beak; therefore it bores holes in living trees and feeds on the sap and the insects attracted to the sap.

Sargasso Fish

Sargasso fish of the Sargasso Sea in the Atlantic Ocean not only change color and design to camouflage themselves, but also have outgrowths resembling weeds, an effective adaptation for avoiding predators in the weed-filled Sargasso.

Scorpion

Like spiders, scorpions are arachnids and have eight legs.

The venom from most scorpions is not fatal to humans, but scorpions of the genus *Androctenus* found in North Africa and members of the genus *Centruroides* of the American southwest are extremely toxic and may be fatal to humans.

Sea Horse

The sea horse is the only fish with a prehensile tail.

Sea Lions and Seals

Seals, sea lions, and walrus are all members of the order Pinnipedia (meaning feather-footed). All members of this order are carnivores adapted for an aquatic existence. The order contains 20 genera and 31 species.

Seals and sea lions occur along the coasts of most of the world and some ascend rivers and live in freshwater lakes. Members of this order are most numerous in polar and temperate waters. The monk seals are the only pennipeds found in tropical oceans.

Seals and sea lions all have torpedo-shaped, streamlined body forms, and all four limbs are modified into flippers. The sea lions are able to rotate their pelvises forward on land to help them walk.

The eyes of the pennipeds have flattened corneas and the pupils are capable of tremendous enlargement as adaptations for seeing effectively underwater.

Seals and sea lions have a thick layer of fat between the skin and the body muscles that serves as insulation and helps them survive in chilly water. In addition, most species have a dense coat of fur.

Seals are known to dive as deep as 2,000 feet and remain submerged for as long as 43 minutes. During deep dives seals use many of the same oxygen-conserving mechanisms as the whales.

Pennipeds have gestation periods of between 8 and 12 months and breed once a year. Delayed implantation of the fertilized egg is common, allowing births to take place at the same time each year.

Seal pups grow rapidly on mother's milk that is approximately 50% fat.

Most species breed communally on beaches,

where bulls compete for breeding territories and establish harems of 3–40 females.

Sea Otter

Sea otters spend the great majority of their lives in the ocean less than a mile off the coasts of California, the Aleutian Islands, and the Kamchatka Peninsula. Sea otters seldom come on land, and when they do, they seldom go more than a few yards from the water.

Sea otters measure up to 5 feet in length and weigh up to 80 pounds.

Sea otters are active during the day and spend the majority of their time floating on their backs. They swim with their bellies down only when they are in a hurry.

Sea otters usually sleep in kelp beds, laying strands of kelp over their bodies to avoid drifting while they are asleep.

Sea otters feed on crabs, mollusks, sea urchins, and fish. They are well known for placing rocks on their chests and using them as anvils to break the shells of their prey. Sea otters consume about one quarter of their body weight in food each day.

Sea otters are born throughout the year and are well developed at birth. Pups nurse for approximately one year, although they take soft foods from their mothers within a few weeks of birth. Pups are carried, nursed, and groomed on their mother's chest.

Sea otters lack the layer of subcutaneous fat found on all other sea mammals and compensate by relying on a layer of air bubbles trapped in their long-haired, extremely dense fur for insulation.

The pelt of the sea otter is generally regarded as one of the most expensive furs in the world. Excessive hunting for pelts drastically reduced the numbers of sea otters until they were extended complete protection in 1911. Today they are increasing in numbers but have only reoccupied about one fifth of their former range.

Secretary Bird

The secretary bird of Africa is a long-legged terrestrial bird of prey. It stands up to four feet tall and has a wingspan of six to seven feet. Its head is adorned with long black-tipped feathers that look like pencils sticking out of a secretary's hairdo.

Secretary birds hunt by walking steadily through short grass savannas in an effort to flush out their prey, primarily rodents, insects, and snakes.

Though normally terrestrial, secretary birds are strong fliers and have been recorded flying at elevations of 12,000 feet.

Shark

The skeletons of sharks are composed entirely of cartilage.

The world's largest living fish is the whale shark, a species that feeds on plankton and measures up to 45 feet in length.

Sharks are extremely ancient fish and have remained virtually unchanged since the age of dinosaurs.

The teeth of sharks grow forward in their jaws and are continually being replaced.

Sheep

There are six species of wild sheep inhabiting North America, the Soviet Union, China, Nepal, Iran, Afghanistan, Cyprus, Kashmir, and Pakistan. Sheep prefer dry, mountainous habitats, often occupying the most rugged country available.

Male sheep have massive, spiraling horns, while those of the females are normally short and only slightly curved.

During the summer males associate in bachelor herds, while ewes congregate with their young. In late fall adult males engage in combat for possession of harems of ewes. During breeding season combat, rams rush together with such force that they are occasionally killed by the impact.

The earliest sheep domestication probably occurred in southwest Asia. The ancestry of the domestic sheep cannot be traced to any known wild species and it is possible that the domestic sheep, classified as *Ovis aries,* is a mixture of several wild species. Bones recognizable as those of domestic sheep have been found in excavations of human settlements dating back to 5000 B.C.

No species of wild sheep has a woolly coat of the type possessed by domestic sheep.

Shrew

There are about 200 species of shrew. They are insectivorous and are among the smallest of all mammals. The Etruscan shrew weighs only grams as an adult.

Shrews are extremely nervous animals. When frightened they may have a heart rate of 1,200 beats per minute, and they have been known to die when frightened by a loud noise.

The salivary glands of some forms of shrew secrete a poisonous substance that they use for subduing small prey and that can cause great pain in humans.

Shrews have an extremely high metabolic rate, and some species consume as much as three times their own body weight daily.

The hero shrews of the African Congo have a remarkable spinal column composed of extremely strong vertebrae. It is known that a 160-pound man can stand on the back of one of these shrews without harming it.

Shrike

The loggerhead shrike, often called the butcherbird, has the habit of impaling its prey, consisting of insects and small rodents, on thorns or barbed wire fences before feeding.

Shrimp

The word *shrimp* comes from the Middle English word *shrimppe* meaning "dried or shriveled."

Red shrimp have been found in the Mariana Trench in the western Pacific as deep as 35,791 feet.

Shrimp swim backward, using their fanshaped tail fins.

Siamese Fighting Fish

The male Siamese fighting fish builds a floating nest for the female by blowing mucus bubbles that bind together at the water's surface. The eggs laid by the female sink to the bottom but are carried to the nest by the male, who binds them in by blowing additional bubbles. The water near the surface contains more oxygen and is a good spawning environment.

Silkworm

The silkworm is the larva of the moth *Bombyx mori*.

One pound of silk comes from 1,500–2,500 silkworm cocoons.

The silkworm moth has been domesticated so long that it is no longer able to fly.

Skua

The skua is a sea gull that frequently obtains food by diving at other birds, forcing them to regurgitate the contents of their crops, then scooping up the regurgitated food before it hits the water. Skuas hunt in more conventional ways too; they often prey on newly hatched penguins.

Skunk

There are approximately ten species of skunks inhabiting the New World. All are characterized by a bold black and white coat pattern.

Skunks are wide-ranging; they live in burrows or almost any other dry place. Skunks are active at dusk and throughout the night, when they emerge to hunt for the small mammals, insects, reptiles, and plant material on which they feed.

In the spotted skunk the white-plumed tail is erected as a warning to potential enemies; if this ploy is unsuccessful, they may stand on their hands and advance in the direction of their enemy.

There is evidence that the hog-nose skunks of South America are immune to the venom of rattlesnakes and other pit vipers. This may also be true of the spotted skunks.

The musk of skunks produces an alarm reaction in rattlesnakes much like the panic that they exhibit in the presence of king snakes. Therefore, rattlesnakes may be an important element in the diets of skunks.

Sloth

Sloths inhabit the tropical forest of central South America. The name *sloth* is a reference to the habitual slow movements of these animals.

A sloth's feet and hands are equipped with long curved claws that they use to hang from three branches, their normal position.

Algae often grows on the sloth's coarse hairs, giving them a greenish appearance, which probably helps them to hide from predators.

The body temperature of sloths, unlike that of most mammals, varies considerably, depending on the surrounding temperature.

Usually slow-moving, sloths move with extreme rapidity when defending themselves, slashing out with their claws and sharp teeth.

Sloths cannot walk on the ground but only drag themselves forward with their claws.

Sloths generally give birth to a single young, which is carried on the chest of the mother, where it clings to her hair.

Snake

The oldest known fossil snakes date to about 65 million years ago and resemble modern boa constrictors and pythons.

Snakes are the most specialized and highly evolved form of reptile. Among the adaptations that all snakes share are an absence of limbs, an elongated body, and eyelids that have joined to form a single, transparent scale, like a contact lens over the eye.

snake

Snakes move very efficiently by producing waves of muscular contractions that produce a side-to-side undulation propelling the snake forward. There are many subtle variations on this pattern of locomotion to allow snakes to climb, move backward, and progress over loose soil.

Snakes are a very successful group of animals and inhabit virtually all available habitats including fresh and salt water; trees; and tropical, arid, temperate, and subterranean environments.

All snakes are carnivorous predators. They swallow their prey whole and are able to eat animals many times larger than their own body diameter because of a flexible joint in the lower jaw. A snake's brain is completely encased in bone to protect it when the snake is swallowing food.

The venom glands of poisonous snakes are actually modified salivary glands. Some snakes developed poison to subdue prey quickly.

The most highly specialized poisonous snakes are the vipers with their hypodermic fangs.

Of the 2,500 species of snake only one quarter are venomous and only about 150 of these are dangerous to humans.

The Paradise tree snake of Southeast Asia eludes predators by launching itself from the tops of trees, spreading its ribs, and gliding into another tree.

Snakes "smell" by means of a specialized organ known as Jacobson's organ. After a snake flicks its tongue out, it inserts the tongue into the Jacobson's organ, which has a large number of sensitive nerve connections leading to the olfactory lobe of the brain.

Sparrow

The English sparrow is actually a type of weaver finch. They have been introduced to many areas of the world and may be the most widely distributed land bird.

Sparrows first appeared in the United States when they were imported to New York City in 1850 to help eliminate tree worms. The original eight pairs of sparrows imported from England died after the first introduction, but more were imported in 1852 and this importation proved successful.

Spiders

All spiders are poisonous, though very few are dangerous to humans.

Spiders are arachnids, as are scorpions, mites, and ticks.

Argyroneta aquatica, a species inhabiting Europe and Asia, builds a bell-shaped web underwater, traps air bubbles with the hairs of its legs, repeatedly transports the air bubbles to the underwater bell until it is filled, then lives in the bell, preying on aquatic species and replenishing its air supply when needed.

The venom produced by the female black widow spider of North America is extremely toxic, yet only about 10% of the people bitten by black widows die because the amount of venom that a black widow can inject is very small.

A tarantula found in Mazatlan, Mexico, in 1935, believed to be 10–12 years of age at the time, lived in captivity for another 16 years.

The largest known spider is *Theraphosa leblondi,* the "bird-eating" spider of South America. This species has a body length of 3.5 inches and a leg span of 10 inches.

Spider silk has one of the highest tensile strengths in proportion to its diameter of any known material, natural or man-made.

Spiders breathe through gill-like structures known as book lungs.

The earliest known spider was *Palaeostenzia crassipes,* which lived 370 million years ago.

There are approximately 40,000 species of spiders.

Sponge

Sponges were considered to be plants by early naturalists because of their limited movement and branchlike growth. The movement to classify them as animals began in the mid-eighteenth century.

Springbok

The springbok, a small antelope measuring about 3 feet at the shoulder and weighing about 75 pounds, is the national animal of South Africa.

Springboks are highly gregarious and at one time migrated across South Africa in herds numbering over a million animals.

Springboks exhibit a behavior called pronking when they are startled or at play. When pronking, the legs are held stiffly, the body is arched, and the head is lowered; in this position the springbok springs repeatedly up to ten feet into the air with no apparent effort.

Squid

Giant squid have the largest eyes of any animal known. Their eyes can be as large as one foot in diameter.

Squid have ten tentacles, of which two are greatly elongated. Squid range in size from one inch to over sixty feet.

Eight of the squid's arms possess two rows of suckers, while the two long arms possess four rows.

Squirrel

Squirrels are found in a wide variety of habitats over most of the world. They do not live in Madagascar,

Australia, southern South America, and the harsh deserts of the Middle East.

Squirrels range in size from the tropical pygmy squirrels weighing less than ten grams to the marmot weighing five or six pounds.

Most squirrels are active during the day, with the exception of flying squirrels, which are nocturnal.

Squirrels normally feed on nuts, seeds, grasses, and other plant material, though some types eat animal food occasionally and a few feed primarily on insects.

Arboreal squirrels are agile in trees and are seldom hurt if they miss their footing and fall. There is a record of a Mexican tree squirrel jumping off a cliff while it was being chased and landing unhurt 600 feet below.

Most tree squirrels have a tendency to glide. When they make flying leaps, they extend their legs, stiffen and curve their tails, and broaden their bodies as far as possible, presenting the broadest surface area and neutralizing the force of gravity to some extent.

The membranes possessed by flying squirrels contain sheets of muscle that allow these specialized squirrels to control the direction of their glides to a certain extent. Just before reaching their landing site, flying squirrels curve their tails up, causing the body to turn up and checking their speed so that they land facing upward on the tree trunk. Flying squirrels have been known to glide as far as 200 feet.

Prairie dogs are ground squirrels. They build and maintain volcano-shaped mounds around their burrow entrances in order to keep the burrow from being flooded during heavy rains.

Starfish

The starfish is the only animal able to turn its stomach inside out. It does this in order to eat its prey, usually mollusks.

The severed arm of a starfish will grow into a completely new and whole starfish, and the original starfish's arm will regenerate.

Stork

Storks are a small family of generally large birds with long legs and necks. They are strong fliers and often soar on thermal air currents. Migrating European white storks for example soar for most of the distance between Africa and Europe.

Storks feed primarily on fish, frogs, large insects, snails, small mammals, and reptiles, which they find by stalking across open plains and marshes. The marabou and adjutant storks feed mainly on carrion, often feeding on the carcasses of large game animals alongside vultures.

In some parts of Africa, marabou storks prey on the eggs and nestlings of pelicans and flamingos.

Storks mate for life and do not breed until they are several years old. Breeding displays include a variety of elaborate postures, dances, and bill clicking.

European storks are considered to be a symbol of good luck. They frequently nest on the chimneys and rooftops of houses in northern Europe.

Sturgeon

Famous for its roe, beluga caviar, the Russian sturgeon is also the largest bony fish in the world. The average female (the larger of the sexes) measures about 7 feet and weighs 335 pounds, but specimens as long as 24 feet and weighing over 3,000 pounds have been recorded.

Swan

Swans are among the largest of flying birds; the mute swan of Europe weighs up to 33 pounds.

Swans are extremely long-lived birds. One mute swan is known to have lived for 102 years.

Swans are tremendously strong and defend themselves with quick snaps of their powerful wings.

Whistling swans have the highest number of individual feathers of any bird, over 25,000, 80% of which are on the head and neck.

The origin of the term "swan song" comes from the fable that a swan, mute during its lifetime, sings a plaintive song when it is preparing to die.

Swift

Swifts are the fastest flying birds, able to attain speeds in excess of 100 mph.

Swifts are closely related to hummingbirds and spend almost their entire lives on the wing, rarely sleeping, and even passing the night while flying. They spend more time awake than any other land bird.

The cave swiftlet of Southeast Asia builds nests on the ceilings of caves composed entirely of their own sticky saliva, which hardens in contact with the air. The cave swiftlet's saliva is extremely high in protein and is the ingredient of bird's-nest soup. In Asia people risk their lives climbing flimsy poles in order to collect the nests of these birds. Up to 3½ million nests have been exported from Borneo in a single year.

Tailorbird

Tailorbirds are a type of warbler inhabiting the Old World. They have the unique habit of actually sewing large leaves together, using their bills and plant fibers to form an envelope in which they construct their nests.

Takin

The takins of Asia are the largest of the goat-antelopes.

The Himalayan takin has a coat golden in color. The golden takin may be the origin of the myth of the ram with the golden fleece.

Tarantula

Tarantulas are among the largest and zoologically oldest and most primitive of the spiders.

Although tarantulas, like all spiders, produce venom, it is not toxic enough to have any effect upon humans despite their fearsome reputation.

In general, tarantulas do not spin webs, the exception being some South American species.

Tasmanian Devil

Tasmanian devils are carnivorous marsupials now confined to the island of Tasmania, though they were also found on the Australian mainland before the introduction of the dingo. Tasmanian devils measure about 3 feet in length and weigh about 15 pounds.

Despite their name and reputation, the aggressive nature of the Tasmanian devil has been greatly exaggerated. Dr. Eric R. Guiler has reported that the majority of devils he has handled were docile and lethargic.

Tasmanian devils are scavengers, consuming fur, viscera, bones, and carrion with their powerful jaws and strong teeth. In addition, they are known to kill small mammals and reptiles, including the extremely venomous black tiger snake.

Tasmanian Wolf

The Tasmanian wolf or thylacine may now be extinct; if not, it is one of the world's rarest mammals. A live thylacine has not been seen since 1933, though evidence of their continued existence is occasionally found.

Thylacines are a remarkable example of parallel

evolution; they are marsupials but look like and occupy the same niche as placental canines.

Thylacines measure up to six feet in length and have a tawny, reddish-brown coat marked with dark stripes on the lower back and tail; they are sometimes called tigers.

Thylacines hunt by tirelessly following their prey until the prey is exhausted, at which time it is killed by quick bites from the wolf's powerful jaws. They are said to feed only on blood and blood-filled organs and only on warm carcasses. What they do not eat is left for scavengers.

Thylacines were relentlessly persecuted by bounty hunters during the nineteenth century and subsequently decimated by an outbreak of distemper early in the twentieth century. They are now fully protected.

Tenrec

The tenrec, an insectivorous animal found on Madagascar and the Comoro Islands, has the record among the mammals for producing the most offspring in one litter: up to 32.

Termite

Termites are five times more apt to strike a house than a fire is.

Termites are not in the ant family, as most people assume, but in the cockroach family.

The compass, or meridianal, termite (also known as magnetic ant), found in Australia, builds nests eight to ten feet high. It builds them so that the narrow ends of the ax-head-shaped nest always face north and south.

An African genus of termite, *Trinervitermes*, build nests that are only about a foot high, but construct underground shafts that go down as far as 130 feet in order to provide access to water.

A queen termite can live as long as fifty years, and in general, termites are especially long-lived insects.

Termites' chief food is the cellulose obtained from dead wood, not always from a house. In forests they serve a valuable function by eating dead trees and stumps and putting back the soil materials that enrich it.

Some species of termites cannot digest wood directly; they have in their intestines tiny protozoa that break down the wood for them.

Some species of termites build clay nests that are so hard they can only be broken by a pickax. Early Spanish settlers in Brazil hollowed out the clay nests and used them as ovens.

Tigers

Tigers are the largest members of the cat family. There are seven races of tiger sparsely scattered throughout Asia. All races are considered to be endangered.

tiger

Largest of all cats, the Siberian tiger of Manchuria and Siberia can measure over 12 feet in length and weigh in excess of 600 pounds.

When hunting, tigers stalk their prey with a silent crawl until within striking distance. They eat animals of all kinds, killing them with their powerful canine teeth. Tigers have been known to prey upon leopards, crocodiles, and other tigers as well as more conventional game.

Tigers are fond of water and are strong swimmers, able to easily cross rivers, lakes, and bays.

It is estimated that a healthy Bengal tiger requires approximately 3 tons of meat per year in the wild, equal to about 70 axis deer.

Despite the fact that tigers are critically endangered in their natural homes, they breed so freely in captivity that some zoos have curtailed their tiger breeding programs.

Like all cats, tigers have rough tongues; this is an adaptation that, along with their relatively short incisor teeth, allows them to scrape small bits of meat off the bones of the animals they feed upon.

Some captive Bengal tigers are white and black, lacking the normal reddish-orange color. All of these animals can trace their origins to one male white tiger named Bohan who was found as a cub in the jungles of India over 30 years ago.

Toad

The Chinese derive medicines from dried toad poison, which, among other things, contains serotonin, a blood vessel constrictor; bufagin; and bufotenine, a hallucinogenic drug.

Toads are voracious eaters. An American toad in captivity ate over 9,000 insects in a three-month period.

Tuatara

The tuatara (*Sphenodon punctatum*), a lizardlike reptile, is found only on twenty small islands in the Cook Strait, off the north island, New Zealand. The tuatara is a relic, the sole survivor of a group of dinosaurs that flourished 150 million years ago.

Tuataras measure 19–29 inches in length and weigh an average of 7 pounds. They have the slowest metabolic rate of any known vertebrate and may live 75–100 years.

Tuataras are the only known reptiles that do not require heat for activity. In the evenings tuataras hunt for worms and crustaceans, their body temperature remaining at around 57°F.

In tuatara eggs, which are buried underground, the multiplication of cells is interrupted during winter, giving the tuatara the longest incubation period of any reptile: 13 to 15 months.

Tuna

The largest species of tuna, the bluefin, can weigh up to 1,500 pounds.

Turkey

The two species of wild turkey are found only in the New World. The common turkey inhabits the eastern United States and Mexico and the ocellated turkey is found from Yucatán to Guatemala.

Turkeys inhabit woodlands and open forests, living in small flocks, feeding on the ground, and roosting in trees at night.

Turkeys feed on grains, seeds, berries, and insects. The common turkey is said to be especially fond of pecans.

Turkeys are polygamous, and during the breeding season males will fight to the death for the possession of females. Nest building, incubation, and care of the young is carried out by the females alone. At the age of about two weeks turkey chicks can fly into low branches to roost for the night.

Domestic turkeys were first brought to Europe by Spanish adventurers, who found them already domesticated by Mexican Indians in the sixteenth century.

Turkeys are the heaviest land birds inhabiting North America. When frightened, however, turkeys can attain a flying speed in excess of 50 mph.

Turtle

The American Society of Ichthyologists and Herpetologists claims that the word *turtle* refers to all reptiles with shells and the word *tortoise* refers only to terrestrial turtles.

The spine of a turtle is fused to its shell.

The oldest known tortoise was a specimen of Marion's tortoise that is known to have lived 152 years and died accidentally.

The alligator snapping turtle has a wormlike growth on the end of its tongue that it uses underwater to lure fish into its massive jaws.

Green turtles travel as far as 1,400 miles to lay their eggs on the same island on which they were hatched.

The largest species of turtle is the Pacific leatherback turtle, whose shell measures about 7 feet in length. Leatherback turtles weigh between 600 and 800 pounds.

Vampire Bat

The vampire bat *(Desmodus rotundus)* inhabits a variety of habitats from northern Mexico southward into Chile and Argentina. As their name implies, vampires feed exclusively on a diet of fresh blood.

An average meal for a vampire bat, a small mammal, is only about one tablespoon of blood.

Vampires alight on the ground and crawl to their intended prey. They make a virtually painless incision with their sharp canine teeth and use their tongues and deeply grooved lower lips to lap blood from the wound. The saliva of the vampire bat contains an anticoagulant that permits the blood to flow freely for as long as feeding takes.

Vampire bats prey heavily on domestic livestock and constitute a danger to these animals and to humans because they can carry and transmit rabies.

Vicuña

The vicuña is one of the two wild cameloids inhabiting South America (the llama and alpaca are domestic forms of the other wild cameloid, the guanaco). They inhabit semiarid grasslands at elevations of up to 18,000 feet in the Andes.

Vicuñas normally travel in herds of 5–15 individuals led by a male. If danger threatens, the male places himself between the source of alarm and the females, allowing them to escape.

Due to their exquisite fur and wool, vicuñas were hunted to near extinction in recent decades but now are afforded complete protection and have made an encouraging comeback in Argentina.

Vulture

Vultures are highly adapted as carrion feeders; the beaks and feet of vultures are relatively weak and incapable of opening the carcass of a recently killed animal. They therefore must wait for the skin to rot or be opened by some other scavenger before they feed.

In locating carrion, sight plays a large role but most scientists believe that smell is an even more important tool in locating carcasses. The discovery that the turkey vulture has the largest olfactory system of any bird seems to confirm this.

New World vultures evolved some 20 million years earlier than the vultures of the Old World, thus the two groups are not closely related. New World vultures were worldwide in their distribution before the evolutionary advent of the Old World vultures.

American king vulture

One species of condorlike bird *(Teratornis mirabilis)* discovered in the deposits of the tar pits at Rancho La Brea, California, was a soaring bird with a wingspan of 16 feet.

The lammergeier, or bearded vulture, an Old World form, has a unique "beard" of stiff, bristlelike feathers under its lower mandible and for some unknown reason colors its white feathered breast with rufous iron oxide.

The Egyptian vulture uses stones which it picks up with its beak and drops to crack the shells of ostrich eggs. It is one of the few types of birds known to use tools.

Walking-Stick

The walking-stick, an insect resembling a twig, is the longest insect in the world, reaching a length of 13 inches.

Walrus

Walrus inhabit open waters of the Arctic Ocean near the edges of the polar ice.

Bulls measure up to 12 feet in length and may weigh as much as 2,900 pounds. The hide may be 2 inches thick and overlie a layer of blubber 6 inches thick.

Walrus feed primarily on shellfish, which they locate and gather with the aid of their tusks, which in males may reach close to 4 feet in length, and with their prominent "mustache," which is composed of about 400 sensitive bristlelike structures.

Eskimos use virtually every part of the walrus for food, clothing, boat building, shelter, oil, or charms. They "fish" for walrus, using strong lines and hooks baited with blubber.

Walrus, like sea lions and unlike the true seals, can move their hind limbs forward under their bodies to help them move across solid surfaces.

Wapiti

The wapiti, or American elk, is the second largest member of the deer family; they are close relatives of the European red deer.

Bull wapiti measure up to 9 feet in length and weigh about 750 pounds. Adult males grow large, magnificent racks of antlers that are efficient defensive weapons and can measure 6 feet in length.

The word *wapiti* is Shawnee, meaning "white deer."

Wasp

Wasps are carnivorous or parasitic. They eat ants, beetles, caterpillars, cicadas, and other insects.

Water Buffalo

Water buffalo *(Bubalus bubalis)* are widely domesticated and used as beasts of burden in the Middle East and throughout Asia. It is thought that true wild water buffalo are to be found in Nepal, Bengal, and Assam, though some authorities believe that these are really feral animals.

Water buffalo have the largest horns of any of the

cattle, measuring up to five feet along the outer edge.

When pestered by insects, water buffalo retreat to the water and submerge themselves until only the nostrils are exposed.

Female water buffalo produce excellent milk. Italians who imported water buffalo into the Po Valley as beasts of burden, first manufactured mozzarella cheese from water buffalo milk.

Weasel

There are about 15 species of weasels inhabiting North and South America, North Africa, Europe, Asia, Java, Sumatra, and Borneo. Weasels are lithe and slender, have short legs, and are strictly carnivorous.

The smallest living carnivore is the dwarf weasel, measuring six inches and weighing less than two ounces.

In winter some species of weasels turn white. This winter fur is known as ermine.

Weasels are generally nocturnal; they hunt small mammals, which they kill with a bite to the base of the skull.

One of the world's rarest mammals is the black-footed ferret (*Mustela nigripes*) of North America. These weasels prey primarily on prairie dogs. A decrease in the number of prairie dogs has probably contributed to the drastic decline in the number of black-footed ferrets.

Whale

Whales are mammals that have adapted to a completely aquatic existence. There are approximately 38 genera and 90 living species of whales, ranging from the small freshwater river dolphins to the small-to medium-sized toothed whales to the blue whale *Balaenoptera musculus*, the largest species of animal ever to inhabit the planet Earth.

Characteristics shared by all whales include tails (flukes), which are set on a horizontal plane rather than a vertical one as in the fish, and nostrils, which open in either a single or double blow hole usually located on the highest point of the head.

Whales do not blow water out of their blow holes; the spout visible in most species is air expelled from the lungs condensing upon contact with the air.

Whales have acute senses of hearing and touch, their eyesight is fair, and they have no sense of smell. Whales produce numerous sounds and probably depend on echolocation for orientation and in searching for food.

The mammary glands of whales have large reservoirs in which milk collects. Females forcefully expel milk into the mouths of their calves by muscular contractions.

The largest toothed whale is the sperm whale *(Physeter catodon)*, which may measure 60 feet in length. These whales have been greatly persecuted by whalers for the high-grade oil they produce.

Sperm whales feed mainly on squid, even attacking giant squid that may measure as long as the whale.

Blue whales may reach a length of over 100 feet and weigh one quarter of a million pounds. (In comparison, the largest dinosaur measured about 90 feet in length and weighed less than 100,000 pounds.) Like all baleen whales, blue whales feed on small shrimplike animals called krill and may consume two to three tons of krill daily.

The killer whale is actually the largest of the dolphins. One killer whale is known to have lived for over 90 years.

Whales are born tail first; if they were born head first, they might drown during delivery. Immediately after birth the cow pushes her calf to the surface to breathe. The calf of a blue whale gains approximately 200 pounds per day.

Wolf

Two species of wolf are recognized: *Canis lupus,* the timber wolf, and *Canis rufus,* the red wolf, now found only in eastern Texas.

Wolves are highly social creatures, living in packs of 4–20 or more animals. The pack hunts cooperatively and is led by the Alpha male and female, who may be the only members of the pack that reproduce, thus ensuring the survival of the litter of pups.

Although wolves bring down large animals such

as elk and moose, much of their food consists of small animals like mice, other rodents, and fish.

Wolves mate for life.

Wolverine

Wolverines, the largest of the weasels, inhabit taiga and forest tundra across the nothern latitudes of the world.

Wolverines are the strongest animals of their size (about 40 inches) and have been known to drive bears and mountain lions from their kills and even kill moose single-handedly when the moose is bogged down by snow.

Wolverine fur is used as trimming around the hoods of cold-weather parkas because it retains less frozen moisture from the breath than any other known fur or synthetic.

Wombat

Unlike other marsupials, wombats have rootless incisor teeth that grow continuously.

Woodchuck

The woodchuck, or groundhog, is the largest member of the squirrel family, *Sciuridae*.

Woodpecker

Rarely seen by humans, the woodpecker's long tongue is barbed and coated with sticky mucus to enable it to lick insects and grubs from the bark of trees.

A woodpecker's beak travels at speeds approaching 100 mph during rapid pecking.

Woodpecker Finch

The woodpecker finch *(Cactospiza pallida)* is one of 13 species of finch inhabiting the Galápagos Islands known as Darwin's finches. Because all of these finches evolved from a single ancestral type of finch on the Galápagos, they were instrumental in Darwin's development of the theory of natural selection to explain how evolution works.

Woodpecker finches use the spines of cactus to pry insect larvae out of holes in tree limbs and from under bark that they are unable to reach with their short bills. They have been known to reuse the same spine over and over again in their pursuit of insects.

Wren

The wren uses its long bill to pierce the eggs of other birds in order to drink the contents.

Yak

The yak is a member of the family Bovidae; it lives in the cold mountainous regions of Asia up to elevations of 20,000 feet.

Wild yak bulls are almost twice the size of domestic yaks, measuring over six feet at the shoulder and weighing over half a ton.

During most of the year bulls roam in groups of two or three, while cows associate in large herds with their calves.

Yaks are sturdy and extremely surefooted. For this reason they have been domesticated for centuries as beasts of burden, as well as for their milk and meat.

Zebra

Zebras inhabit eastern, central, and southern Africa, preferring plains and savannas, with the exception of Hartmann's Mountain zebra, which is found in mountainous terrain.

Grant's zebra

Grant's zebra

The three living species of zebra are easily distinguished from one another by such characteristics as body and ear size and striping patterns.

Zebras normally live in herds of 10–30 individuals led by a stallion.

Zebras are resistant to some common African diseases that domestic horses cannot survive. Attempts have been made to domesticate zebras for this reason, but to date they have been unsuccessful.

Male zebras have large pointed canine teeth that they use in defense and in battles with other males for the possession of females.

The word *zebra* comes from the Amharic-derived Portuguese word *zebra,* meaning "wild ass."

Romans called zebras "hippotigris," or striped horse.

The largest species of zebra is Grévy's zebra *(Equus grevyi),* named after a past president of France, Jules Grévy, which measures almost 5 feet at the shoulder and weighs about 750 pounds.